FRAME RELAY NETWORKING
Covers all aspects of Frame Relay Networking, from the evolution and rationale to architecture and equipment.
1999 0 471 98578 3

INTERNETWORKING LANs AND WANs
Concepts, Techniques and Methods
Second Edition

Internetworking is one of the fastest growing markets in the field of computer communications. However, the interconnection of LANs and WANs tends to cause significant technological and administrative difficulties. This updated version provides valuable guidance, enabling the reader to avoid the pitfalls and achieve successful connection.
1998 0 471 97514 1

THE MULTIPLEXER REFERENCE MANUAL

Designed to provide the reader with a detailed insight into the operation, utilization and networking of six distinct types of multiplexers, this book will appeal to practising electrical, electronic and communications engineers, students in electronics, network analysts and designers.
1993 0 471 93484 4

PRACTICAL NETWORK DESIGN TECHNIQUES

Many network design problems are addressed and solved in this informative volume. Gil Held confronts a range of issues including through-put problems, line facilities, economic trade-offs and multiplexers. Readers are also shown how to determine the numbers of ports, dial-in lines and channels to install on communications equipment in order to provide a defined level of service.
1991 0 471 92938 7 (Set)

HIGH SPEED
DIGITAL
TRANSMISSION
NETWORKING
SECOND EDITION

HIGH SPEED DIGITAL TRANSMISSION NETWORKING

COVERING T/E-CARRIER MULTIPLEXING, SONET AND SDH

SECOND EDITION

Gilbert Held

4-Degree Consulting,
Macon, Georgia, USA

JOHN WILEY & SONS

Chichester • New York • Weinheim • Brisbane • Singapore • Toronto

First edition published in 1990 as Digital Networking and T-Carrier Multiplexing
Copyright © 1999 by John Wiley & Sons Ltd
    Baffins Lane, Chichester,
    West Sussex, PO19 1UD, England

    National 01243 779777
    International (+44) 1243 779777
e-mail (for orders and customer service enquiries): cs-books@wiley.co.uk

Visit our Home Page on http://www.wiley.co.uk or http://www.wiley.com

Other Wiley Editorial Offices

John Wiley & Sons, Inc., 605 Third Avenue,
New York, NY 10158-0012, USA

WILEY-VCH Verlag GmbH
Pappelallee 3, D-69469 Weinheim, Germany

Jacaranda Wiley Ltd, 33 Park Road, Milton,
Queensland 4064, Australia

John Wiley & Sons (Canada) Ltd, 22 Worcester Road,
Rexdale, Ontario, M9W 1L1, Canada

John Wiley & Sons (Asia) Pte Ltd, 2 Clementi Loop #02-01,
Jin Xing Distripark, Singapore 129809

Library of Congress Cataloging-in-Publication Data

Held, Gilbert 1943–
    High speed digital transmission networking. : covering T/E-carrier
multiplexing, SONET and SDH/Gilbert Held.
        p. cm.
    Includes index.
    ISBN 0-471-98358-6 (alk. paper)
    1. Data transmission systems. 2. Digital communications.
3. Multiplexing. 4. SONET (Data transmission) 5. Synchronous
digital hierarchy (Data transmission) I. Title.
TK5105.H4275 1999
621.382—dc21   98-54675
                                                         CIP

British Library Cataloguing in Publication Data

A catalogue record for this book is available from the British Library

ISBN 0 471 98358 6

Typeset in 10/12.5pt Bookman Light by Dobbie Typesetting Ltd, Tavistock, Devon
Printed and bound in Great Britain by Bookcraft (Bath) Ltd
This book is printed on acid-free paper responsibly manufactured from sustainable forestry, in which at least two trees are planted for each one used for paper production.

CONTENTS

6   T- and E-carrier framing and coding formats   133

9   Testing and troubleshooting   223

PREFACE

This book is focused upon the current and evolving communications carrier digital transmission infrastructure and third party equipment which enables business, industry and academia to transmit voice, data and video at high speed. This communications infrastructure can begin at your office or residence and consists of copper or fiber cable routed tens or thousands of kilometers to the party you wish to communicate with. By focusing upon the existing and evolving communications carrier digital transmission infrastructure, you can obtain an appreciation for the use of different data transport facilities as well as understand how to consider such important issues as their reliability, availability, utilization economics and practicality of use to access various types of packet switching services. Those services can include X.25 and Frame Relay networks and the Internet. In addition, you will obtain a mechanism for effectively and efficiently constructing private organizational networks through the selection of different types of digital transmission facilities.

Digital transmission facilities covered in this book range in scope from Dataphone Digital Service and fractional and full T- and E-carrier systems to different optical systems. By understanding how such systems operate and the advantages and disadvantages associated with their use, you can match an appropriate transmission facility to your organization's communications requirements.

Most communications-related books, including those previously written by this author, are primarily focused upon a particular service or protocol such as Frame Relay, TCP/IP and Internet-related topics. This book is different in that it examines communications from the data transport facility viewpoint, focusing upon different copper- and fiber-based transport mechanisms installed by communications carriers which permit access to different service offerings and also enable you to construct an organizational private network. As such, this book provides you with information that can be used to select an appropriate transport facility by understanding the reliability, availability, testing and economic issues associated with each type of transport facility. Since transmission facilities by themselves only represent a mechanism for bits to flow on copper or fiber, this book is also focused upon equipment necessary to transport information effectively and efficiently over different types of communications carrier infrastructures. To accomplish this objective, this book examines the use of T- and E-carrier multiplexers, digital modems and other devices which provide a mechanism for transmitting voice, data and video. Since one of the more popular applications of high speed transmission is the movement of voice and data over a common transmission facility, this book also examines many popular voice digitization methods, as an understanding of those methods is integral to the effective and efficient use of different types of equipment on high speed transmission facilities. By focusing on the communications carrier infrastructure and equipment to move information over that infrastructure, you can obtain information necessary to perform high speed networking. Since that infrastructure is based upon digital technology, the title of this book represents its focus upon the digital infrastructure and equipment necessary to transmit voice, data and video at high speed over that infrastructure.

In this book we will focus our attention upon high speed digital networking, including the advantages and disadvantages associated with digital signaling, the operation and utilization of communications carrier offerings, the frame formats of North American and European T- and E-carriers, the features and operational utilization of T- and E-carrier multiplexers, and the methods by which digital facilities can be tested. Each of these areas is investigated in a hierarchical manner, with each chapter building upon the material presented in the preceding chapter. It is, therefore, recommended that readers unfamiliar with digital networks and digital signaling techniques should read each chapter in sequence. More experienced readers requiring specific information, such as testing and troubleshooting methods, can consider beginning their reading at a specific chapter in this book.

In developing the material incorporated into this book I have structured it as a classroom text for students who have previously completed an introductory course in data communications. In addition, this book was also written for the data communications practitioner working in industry or government who has equivalent experience through on-the-job training and the reading of appropriate introductory books, college courses or seminar attendance. For both categories of readers, without appearing to be self-serving, I would recommend my previous book, *Understanding Data Communications: From Fundamentals Through Networking* (2nd edition), which provides a comprehensive introduction to the field of data communications.

Since the present book was written primarily as an advanced text for students completing an introductory course in data communications, it should be considered as a one-semester course for use by seniors or first-year graduate students majoring in telecommunications or computer science. To assist both students and practitioners I have included a series of review questions at the end of each chapter. These questions were selected to reinforce key concepts, and readers are encouraged to work each problem prior to going forward in the book.

As a professional author who has worked in the communications technology field for over 20 years, I sincerely welcome reader feedback. Your comments concerning the content and flow of material presented, the need for additional information and other matters are most welcome. You can reach me through my publisher whose address is given at the front of this book or via electronic mail at 235-8068@mcimail.com.

Gilbert Held
Macon, GA

ACKNOWLEDGMENTS

Although this book bears my name as author, its evolution from a draft manuscript through the various stages of production represents a team effort. Thus, I would be remiss if I did not take the opportunity to acknowledge the efforts of the many people who contributed to it.

The fact that a book requires a publisher is self-evident. However, in the technical publishing area the publisher must be able to understand the evolving communications field to decide upon the viability of proposals. Once again, I am indebted to Ann-Marie Halligan and her staff for arranging for the reviews of my initial proposal, providing guidance and backing this writing project.

As an old-fashioned author I create draft manuscripts using pen and paper. This enables me to work on writing projects without worrying about battery levels, different outlet receptacles and restrictions on the use of electronic apparatus on airlines as I travel around the globe. Converting my pen and paper drafts into an electronic manuscript, including creating professional drawings from my hand-generated illustrations, is a major effort. Thus, I am most grateful for the effort of Mrs Linda Hayes who once again carried out this task with her usual efficiency and professionalism.

Once a manuscript is completed, the publishing effort is focused on the various stages of the production process required to produce the book. This is a considerable task which ranges from the copy editing of the manuscript all the way through the typesetting and cover design to the printing and binding of the book. I would like to take the opportunity to thank Juliet Booker, Robert Hambrook, Sarah Lock and the rest of the production department at John Wiley & Sons for their fine work.

Last but not least, I would like to thank my family for the lost evenings and weekends during which I hibernated in my office researching and drafting the manuscript for this book. I truly appreciate their understanding, cooperation and assistance.

1

RATIONALE FOR HIGH SPEED DIGITAL NETWORKING

Today we live in an information-based society, with the strength of nations more dependent upon their communications infrastructure than on their production of steel. This change to an information-based society did not happen overnight. Instead, it represents an evolution of the communications carrier infrastructure and the development of a variety of communications equipment over a period of approximately half a century. During that period it is important to remember that at one time 110 bps and 300 bps modems were considered state-of-the-art high speed transmission devices. Of course, when 110 bps and 300 bps transmission was considered state-of-the-art there was no Internet, LANs did not exist, and the use of email to transmit images taken through the use of digital cameras was found only in a science fiction novel.

When we use the term 'high speed digital networking' it is common for different persons to have different expectations concerning the term high speed. For some, 4800 and 9600 bps could represent high speed, especially if their use of communications is to display information on a terminal screen and read the displayed information. For others 1.544 Mbps, which is the T1 operating rate, could represent a slow data rate. Recognizing this potential diversity in the viewing of what is a high speed digital transmission facility, I will attempt to examine the operation and utilization of most digital transmission facilities and let you decide which is an appropriate facility to meet your specific networking requirements. However, the primary focus of this book follows its title, with special emphasis placed upon digital transmission facilities that operate at and above 1.544 Mbps, the operating rate considered to represent a high speed transmission capability by this author.

In this chapter we will examine why both large and small business organizations, communications carriers, government agencies and universities are migrating their voice and data communications to high speed digital networks. Although there are many key advantages obtained from digital networking, as might be expected from the use of any technology, there are also disadvantages associated with the use of this technology.

To obtain an understanding of the advantages and disadvantages of high speed digital networking first requires a comparison of the method of operation of analog and digital transmission systems. Once this comparison is completed we will use this information as a base to explore the advantages and disadvantages of digital signaling which forms the basis for the construction of digital networks. This information will provide us with a significant indication of the rationale behind the large investments both communications carriers and end-user organizations are making to obtain a high speed digital networking capability.

After this has been accomplished, we will turn our attention to bandwidth consumption issues, examining the potential effect of modern applications upon the capacity of different types of transmission facilities. In doing so we will obtain an appreciation for the reason why high speed digital networks provide a mechanism for satisfying the requirements of modern bandwidth-consuming applications.

1.1 ANALOG TRANSMISSION

In analog transmission a continuous wave is modulated to impress information onto a carrier signal. The resulting modulated signal is subject to several types of impairments as it propagates down a transmission path, including attenuation and envelope delay.

The resistance, capacitance and inductance of a circuit attenuate the signal, resulting in a reduction in its signal strength. In addition, the construction of communications carrier analog voice-grade facilities is such that only frequencies between 300 Hz and 3300 Hz are passed, forming the voice-grade passband illustrated in Figure 1.1. This passband, which is a contiguous portion of an area in the frequency spectrum, is formed by the use of low-pass and high-pass filters and results in a bending of the attenuation-

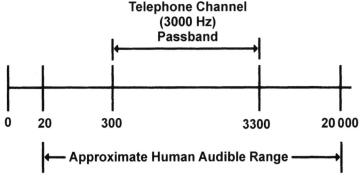

Figure 1.1 Passband of a telephone channel. The telephone channel passband is a contiguous area of the frequency spectrum formed by the use of filters which permits frequencies between 300 and 3300 Hz to pass

frequency response of an analog signal as it propagates along the transmission path. Since high frequencies attenuate more rapidly than low frequencies, this natural phenomenon results in a further distortion of the received signal.

A second problem associated with analog transmission is caused by the delay of certain frequencies of a composite signal as they flow through a communications channel. Usually, the filters employed by the communications carrier delay frequencies near the cutoff frequency of the filters. In addition, the non-linear relationship between the phase shift and frequency of analog signals results in the propagation of the harmonics of the signal at different velocities. Together, the filters used by the communications carriers and the difference in the velocity of propagation of the harmonics of a complex signal result in an envelope delay curve similar to that illustrated in Figure 1.2.

Amplifiers and equalizers

At periodic intervals throughout their initially developed networks, communications carriers installed amplifiers to compensate for the loss in strength of analog signals. Although an amplifier will boost signal strength, it also increases any noise or distortion that previously occurred to the signal. This is illustrated in Figure 1.3. Thus, after amplification, the effect of noise and distortion that may have previously occurred is also increased.

To minimize the effect of the amplitude–frequency response of a signal and any envelope distortion, analog leased lines can be conditioned. What is known as C-level conditioning in the United States results in the communications carrier installing equalizers

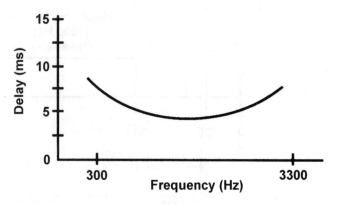

Figure 1.2 Envelope delay curve. Both the filters used by a communications carrier and the difference in the velocity of the harmonics of a complex signal result in some portions of a signal arriving at its destination prior to other parts of the signal. This delay is known as envelope delay

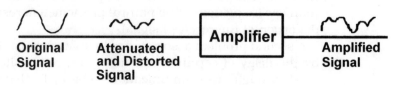

Figure 1.3 Amplification of an analog signal. Although an amplifier boosts the strength of an analog signal, it also increases the effect of any noise and distortion that occurred to the signal

to generate both an amplitude–frequency response and a delay inverse to that occurring to a signal traversing the circuit. The resulting objective of the use of equalizers is to produce a uniform delay and uniform amplitude–frequency response over the pass-band.

Figure 1.4 illustrates the theoretical effect obtained by the use of attenuation equalizers, while Figure 1.5 illustrates that obtained by the use of delay equalizers. As illustrated in Figure 1.4, the attenuation equalizer introduces a variable gain inverse to the attenuation loss occurring on the circuit, theoretically resulting in a uniform attenuation across the frequency spectrum. Similarly, the use of delay equalizers is designed to provide a uniform delay to all components of a signal over the voice channel passband. This is accomplished by the delay equalizer introducing a variable delay inverse to the actual delay occurring on the circuit.

In actuality, the equalizers used by communications carriers are fixed and are set at circuit installation time. Since line conditions vary, the equalizers are used to provide a range of acceptable

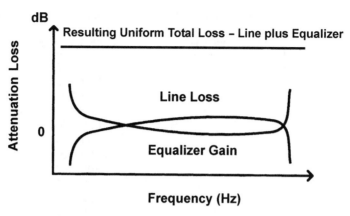

Figure 1.4 Using an attenuation equalizer. An attenuation equalizer introduces a variable gain inverse to the line loss in an attempt to obtain a uniform total loss

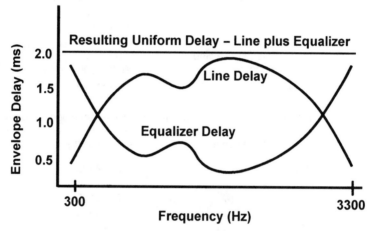

Figure 1.5 Using a delay equalizer. A delay equalizer introduces a variable delay inverse to the line delay in an attempt to obtain a uniform total delay

values that the carrier attempts to provide, based upon the level of conditioning selected. Tables 1.1 and 1.2 list the bandwidth parameter limits for frequency loss in decibels (dB) and envelope distortion delay in microseconds (μs), respectively. The BASIC entry references an unconditioned channel, while C1, C2, C4 and C5 represent four types of conditioning available from AT&T. British Telecom Keyline bandwidth parameter limits are the same as the International Telecommunications Union (ITU) (formerly known as the Consultative Committee for International Telephone and Telegraph (CCITT)) M. 1020 four-wire international leased line recommendation. Note that a negative dB limit in Table 1.1 means less loss or a gain while a positive dB limit is more loss or an actual signal loss.

Table 1.1 Bandwidth parameter limits — frequency response loss.

Channel/conditioning	Frequency range	Limits (dB)
BASIC	300–500	−3 to +12
	500–2500	−2 to +8
	2500–3000	−3 to +12
C1	300–1000	−3 to +12
	1000–2400	−1 to +3
	2400–2700	−2 to +6
	2700–3000	−3 to +12
C2	300–500	−2 to +6
	500–2800	−1 to +3
	2800–3000	−2 to +6
C4	300–500	−2 to +6
	500–3000	−2 to +3
	3000–3200	−2 to +6
C5	300–500	−1 to +3
	500–2800	−0.5 to +1.5
	2800–3000	−1 to +3
British Telecom Keyline	300–500	−2 to +6
(ITU M.1020)	500–2800	−1 to +3
	2800–3000	−2 to +6

Table 1.2 Bandwidth parameter limits — envelope distortion delay.

Channel/conditioning	Frequency range	Limits (μs)
BASIC	800–2600	1750
C1	800–1000	1750
	1000–2400	1000
	2400–2600	1750
C2	500–600	3000
	600–1000	1500
	1000–2600	500
	2600–2800	3000
C4	500–600	3000
	600–800	1500
	800–1000	500
	1000–2600	300
	2600–2800	500
	2800–3000	3000
C5	500–600	600
	600–1000	300
	1000–2600	100
	2600–2800	600
British Telecom Keyline	500–600	3000
(ITU M.1020)	600–1000	1500
	1000–2600	500
	2600–3000	3000

To support high-speed transmission on the switched telephone network and on leased lines, modem manufacturers included automatic and adaptive equalization circuitry in their products. Each time the direction of data flow changes, one modem sends a 'training' signal which is used by the modem to adjust its level of equalization. This process of automatic and adaptive equalization affects data throughput as well as adds to the complexity and expense of the modem.

Another problem associated with the use of automatic and adaptive equalizers is the training time that can be required after a line hit. Some modems can take as long as several seconds to retrain after a line hit, resulting in the absence of data being transmitted during that time. If a time-dependent protocol is being used when a line hit occurs, the modem retraining time can result in a timeout, causing the protocol to go into a reinitialization routine which further reduces transmission throughput.

Although almost 100% of the backbone network infrastructure of most communications carriers was converted to digital transmission by the late 1990s, the copper wire connection between most subscribers and the carrier's backbone digital network is primarily used to transport data in an analog format. Thus, the previously described characteristics of analog transmission will adversely effect your ability to transfer data unless you obtain an end-to-end digital transmission capacity. As we will note later in this book, many communications carriers will install fiber or terminate a copper-based digital circuit within a building to provide an end-to-end digital transmission capability.

1.2 DIGITAL TRANSMISSION

When digital transmission facilities are used, data travels from end to end in a digital format. Although digital pulses are subject to impairments similar to those experienced by analog signals, carrier facilities employ repeaters instead of amplifiers on copper transmission facilities, which significantly reduces the effect of distortion upon the signal.

Use of repeaters

At regular intervals on a digital transmission facility, the communications carrier installs repeaters whose function is to rebuild or regenerate pulses into their original form.

Figure 1.6 illustrates the operational effect from the use of a repeater in a digital transmission system. The repeater scans the

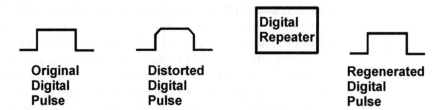

Figure 1.6 Repeaters eliminate distortion. When digital transmission facilities are used, data travels from end to end in a digital format. Digital pulses are regenerated at regular intervals which results in the elimination of distortion

line looking for a pulse rise and then discards the rest of the pulse, regenerating a new digital pulse in place of the incoming pulse. This operation results in the elimination of any prior distortion and explains why, in general, a digital transmission system is more reliable than an analog transmission system that employs amplifiers which boost both the signal and any prior distortion the signal received.

Repeaters, which are also referred to as data regenerators, are used on both copper and fiber optic transmission facilities. Copper repeaters operate upon voltage while fiber optic repeaters base their pulse regeneration effort upon light intensity. Both types of repeaters eliminate distortion by regenerating either a new electrical or a new optical pulse.

1.3 ADVANTAGES AND DISADVANTAGES OF DIGITAL SIGNALING

Digital networks are based upon the use of digital signaling. Thus, we can examine the basis for the growth in digital networking by examining the advantages and disadvantages associated with digital signaling.

Table 1.3 lists the primary advantages and disadvantages of digital signaling in comparison to analog signaling. By reviewing each of the entries in the table we can obtain an appreciation behind the rationale for the growth in the use of digital transmission facilities as well as some of the problems that can be expected to occur when using this technology.

Advantages

Reliability

In terms of reliability, a digital network can be expected to have a lower error rate than an analog network. This expectation is based

Table 1.3 Advantages and disadvantages of digital sampling.

Advantages	Reliability
	Multiplexing efficiency
	Voice and data integration capability
	System performance monitoring capability
	Ease of encryption
	Economics
	Path to ISDN, ATM, SDH and SONET
Disadvantages	Loss of precision in analog to digital conversion
	Analog system interface requirements
	Increased bandwidth requirements
	Need for precise timing

upon the previously discussed differences between amplifiers and repeaters with respect to an analog signal being boosted to include any noise and distortion, while a digital pulse is discarded by the repeater which then generates a completely new pulse. A second significant difference between analog and digital signaling reliability results from the equalizers required for high speed analog transmission. Although equalizers are built into digital service units (DSUs) which are the digital equivalent of a modem, their circuitry is far less complex than equivalent circuitry incorporated into modems. In addition to DSUs having a cost significantly less than that of a high speed modem, their lower complexity makes them more reliable. Finally, the retraining time on a digital network is usually significantly less than the retaining time of modems used on analog networks, in effect significantly reducing the possibility of a timeout occurring.

Multiplexing efficiency

When comparing the ease of multiplexing analog to digital signals, we must compare the technology associated with each technique. In frequency division multiplexing (FDM), tuned filters are required to derive subchannels from the available frequency spectrum of a circuit. To prevent frequency drift resulting in the modulated data on one subchannel interfering with the transmission of data on adjacent subchannels, guard bands are assigned between subchannels. Figure 1.7 illustrates the splitting of a 3 kHz voice channel into subchannels, with guard bands employed to minimize the effect of frequency drift between channels. The guard bands are unused segments of frequency which alleviate the effect of

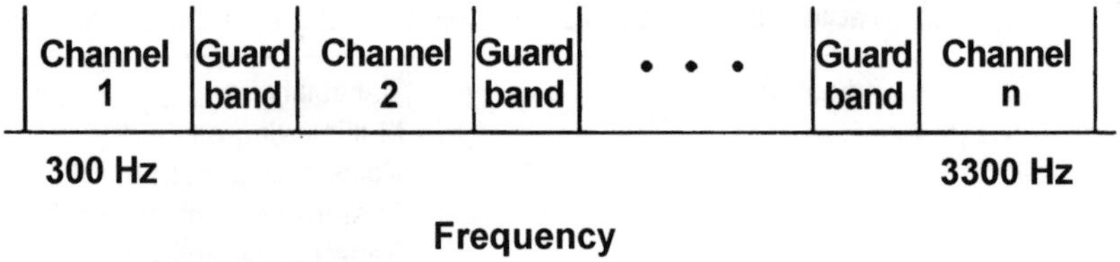

Channel 1	Guard band	Channel 2	Guard band	• • •	Guard band	Channel n

300 Hz   3300 Hz

Frequency

Figure 1.7 FDM channel separations. In frequency division multiplexing, the 3 kHz bandwidth of a voice-grade line is split into channels or data bands separated from each other by guard bands

Table 1.4 FDM channel spacings.

Speed (bps)	Spacing (Hz)
75	120
110	170
150	240
300	480
450	720
600	960
1200	1800

frequency drift associated with analog devices. Unfortunately, the use of guard bands also removes portions of the frequency spectrum that could be used for transmission, reducing the overall efficiency of frequency division multiplexing. A second limitation associated with FDM equipment is the channel spacings required for multiplexing different data rates. As indicated in Table 1.4, FDM channel spacing requirements would not enable two 1200 bps data sources to be multiplexed on a conventional voice-grade channel, since only 3000 Hz is available for use and each 1200 bps source would require 1800 Hz.

The use of time division multiplexers (TDM) to support digital transmission can result in a higher level of efficiency. This is because the aggregate data transmission capacity of the composite channel is the limiting constraint and the multiplexer interleaves data from low to medium speed sources onto the high speed channel by time.

In comparing FDM to TDM you must also consider the reliability of equipment and its ease of modification. In both cases, TDM equipment holds an advantage over FDM equipment, since digital components are more reliable than analog components and it is easier to adjust the assignment of time slots than to vary subchannel separations that are controlled by the use of tuned filters.

Voice and data integration capability

Since the early 1960s, communications carriers have been adding equipment and modifying their network facilities to support digital signaling. At carrier central offices, equipment has been added that digitizes voice conversations, enabling both voice and data to be multiplexed and transmitted together on trunk circuits connecting carrier offices.

The success experienced by communications carriers in the integration of voice and data eventually resulted in the commercial offering of high speed communications facilities known as T- and E-carrier circuits. This, in turn, resulted in equipment vendors developing products that enabled end-user organizations to design their own integrated voice/data networks using T- and E-carrier facilities. Initially T- and E-carrier transmission facilities were installed based upon the use of copper wiring between subscribers and the communications carrier serving the central office. Commencing during the 1980s, the transmission requirements of many organizations increased to the point where the installation of fiber directly into a building became economically feasible. With the development of the Synchronous Optical Network (SONET) and Synchronous Digital Hierarchy (SDH) network, it is quite common for organizations with high speed transmission requirements that also require a high degree of transmission reliability to have fiber optic rings installed between the serving office of a communications carrier and their building. The fiber ring, operating as either a SONET or SDH transmission facility, provides a high speed optical transmission capability that can support the operation of multiple T- or E-carrier subchannels. In addition, the use of fiber ring technology adds a significant level of redundancy as many rings can be installed with a self-healing feature. This allows the opposite flow of data to occur in under a millionth of a second, which restores communications if a user's activities result in a break in the ring.

The operation and utilization of T- and E-carrier facilities, as well as the use of multiplexers containing different types of voice digitization cards, and the operation of fiber optic rings and their self-healing capabilities, are described later in this book.

System performance monitoring capability

In an analog transmission system the measurement of circuit performance requires the insertion of a known pattern of test bits which interrupts the flow of data. In comparison, two framing formats used for T-carrier transmission include a mechanism whereby performance data can be obtained by monitoring the circuit without interrupting the flow of data. On an E-carrier system, as well as on many fiber optic-based systems, separate channels are defined within a multiplexing hierarchy, which enables testing to occur and performance measurements to be made without interrupting the flow of data.

A second advantage of digital signaling with respect to analog signaling is the coding mechanism used to place digital data onto a digital transmission facility. This coding mechanism results in the ability of equipment to identify errors without requiring the equipment to have knowledge of the information that is being transmitted. These errors are detected as bipolar violations (BPV), and both digital encoding techniques and bipolar violations are discussed in detail later in this book.

Ease of encryption

Although analog transmission can be encrypted, to do so is both difficult and costly. In addition, its result may be reversible within a short period of time. This is because the encryption of an analog signal requires the use of numerous filters to divide the frequency spectrum into subchannels. The resulting subchannels are then moved to different positions within the frequency spectrum, in effect making a telephone conversation indecipherable to a person monitoring the transmission.

When digital signaling is used, encryption is both more secure as well as easier to implement. This is because a pseudo-random bit string can be generated by a key and added via modulo-2 addition to the source digital data stream to encrypt transmission. Then the ability to change the key used to generate the pseudo-random bit string allows almost an infinite number of codes to be generated. In comparison, the bandwidth of an analog voice-grade line and the physical constraints in developing filters and circuitry necessary to scramble a voice conversation results in a relatively low number of channel positioning possibilities in comparison to the number of pseudo-random bit strings that can be generated using digital circuitry.

Economics

On a cost per bit per second (bps) basis, digital transmission facilities are in certain situations much more economical than analog facilities.

Table 1.5 Interoffice channel cost, cost per mile in US dollars (1 mile=1.6 km).

Facility	Cost per mile
Analog circuit	0.28
2.4 kbps	0.28
4.8 kbps	0.28
9.6 kbps	0.28
56.0 kbps	0.68
T1-carrier (1.536 Mbps effective)	4.06
T3-carrier (43.008 Mbps effective)	59.88

The maximum data rate obtainable on analog leased lines was 33.6 kbps using trellis coded modulation modems that were relatively expensive when this book was written. In comparison, digital transmission facilities can be obtained to provide a transmission rate of 2.4 to 56 or 64 kbps for the use of subrate facilities and 1.544/2.048 Mbps on T- and E-carriers by the use of inexpensive digital service units. Here 1.544 Mbps is the T-carrier data transmission rate in North America while the 2.048 Mbps data rate is used in Europe, which is also known as E1 transmission.

Another popular T-carrier transmission facility we can use for discussing economics is the T3 circuit. Although this circuit operates at approximately 45 Mbps, it represents an aggregation of 28 T1 circuits and has a significant amount of overhead. Thus, when we turn our attention to comparing the cost of analog and digital transmission facilities, we will use their actual data transfer capability which will provide a more realistic comparison.

Although there are numerous factors that contribute to the precise rate paid for analog and digital circuits, including access connection fees, variable interoffice channel charges based upon the distance between locations to be connected, and installation cost, for simplicity we will focus our attention upon mileage cost. For illustrative purposes, Table 1.5 lists the approximate costs per mile (1 mile=1.6 km) for an analog leased line and several types of digital transmission facilities that were in effect by one communications carrier when this book was written. Obviously, for an up-to-date economic comparison, you should contact one or more communications carriers to obtain both the current interoffice channel cost and other fees appropriate to each type of circuit under consideration.

For the cost-effectiveness examples that follow we will assume that the cost of a 33.6 kbps modem is $200 per unit. In addition, let us assume that the cost of a DSU operating at data rates up to 9.6 kbps is $300 while the cost of a 56 kbps DSU is $400. For a T-carrier circuit operating at 1.544 Mbps, the DSU function is normally built into multiplexing equipment, and another device known as a channel service unit (CSU) is used for transmission on that type of circuit. Thus, we will assume that the cost of a T1-carrier CSU is $850 in the economic comparisons we will perform, while we will use a cost of $4000 for a T3 CSU. Finally, in each of the following economic comparisons we will assume that purchased equipment, such as modems or DSUs and CSUs, are amortized over a three-year (36-month) period.

Thus, the cost of a 500-mile analog circuit operating at 33.6 kbps on a per bps per month basis becomes

$$\frac{(2 * \$200/36 \text{ months}) + 0.28/\text{mile} * 500 \text{ miles}}{33.6 \text{ kbps}}$$

$$= 0.00449 \text{ per bps per month}$$

Now let us examine the cost of several types of digital transmission facilities. The cost of a 500-mile digital facility operating at 2.4 kbps on a per bps per month basis is

$$\frac{(2 * \$300/36 \text{ months}) + 0.28/\text{mile} * 500 \text{ miles}}{2.4 \text{ kbps}}$$

$$= 0.0653 \text{ per bps per month}$$

Using the same method of cost computation, the monthly cost per bps for a 500-mile digital circuit operating at 4.8 kbps becomes

$$\frac{(2 * \$300/36 \text{ months}) + 0.28/\text{mile} * 500 \text{ miles}}{4.8 \text{ kbps}}$$

$$= 0.0326 \text{ per bps per month}$$

Now let us compute the cost of a 500-mile digital circuit operating at 9.6 kbps on a per bps per month basis. This cost is computed as follows:

$$\frac{(2 * \$300/36 \text{ months}) + 0.28/\text{mile} * 500 \text{ miles}}{9.6 \text{ kbps}}$$

$$= 0.0163 \text{ per bps per month}$$

Note from the preceding analysis that the cost of a very low-speed digital transmission facility is normally more expensive than an analog circuit on a per bit per month basis. While there are many other factors that should be investigated, a digital transmission facility does not always imply an economic saving over an analog facility. Now let us examine the cost of three additional digital facilities. The cost of a 500-mile 56 kbps digital facility on a per bps per month basis is

$$\frac{(2 * \$400/36 \text{ months}) + 0.68/\text{mile} * 500 \text{ miles}}{56.0 \text{ kbps}}$$

$$= 0.0065 \text{ per bps per month}$$

The cost of a 500-mile 1.544 Mbps T-carrier circuit on a per bps per month basis is

$$\frac{(2 * \$850/36 \text{ months}) + 4.06/\text{mile} * 500 \text{ miles}}{1.536 \text{ kbps}}$$

$$= 0.0014 \text{ per bps per month}$$

Note that in the preceding computation a data rate of 1.536 Mbps was used as the divisor to obtain the monthly cost per bps. The 1.536 Mbps data rate was used because 8 kbps on a T1 line is used for framing and cannot transport data. Thus, 1.544 Mbps minus 8 kbps results in an effective data rate of 1.536 Mbps.

Since a T3 circuit represents an aggregation of 28 T1 circuits, its effective data rate becomes 1.536 Mbps * 28 or 43.008 Mbps. Thus, the cost of a 500-mile T3 carrier circuit on a per bps per month basis is

$$\frac{(2 * \$4000/36 \text{ months}) + 59.88/\text{mile} * 500 \text{ miles}}{43.008 \text{ Mbps}}$$

$$= 0.0007 \text{ per bps per month}$$

Table 1.6 summarizes the previously computed costs of the voice-grade analog circuit and the six digital circuits on a per bps per month basis. Note that only when a digital transmission facility operates at or above 56 kbps is the cost per bps per month comparable with that of an analog transmission facility. In fact, the cost of a 1.544 Mbps T-carrier digital facility is approximately one-third that of an analog voice-grade circuit on a bps per month basis.

A second method of comparing the cost of analog and digital transmission facilities involves examining the voice and data carrying capacity of a T1-carrier circuit. In North America, a T1-carrier can be used to transmit data, a mixture of digitized voice and data, or it can be used to transmit at least 24 digitized voice conversations. If we assume the latter, the economics of using a T1-carrier facility become substantially more pronounced. As an example of this, in early 1999 the cost of an analog voice-grade circuit between Macon, Georgia, and Washington, DC—a distance of about 650 miles—was approximately $182 per month. In comparison, a T1-carrier circuit capable of supporting 24 digitized voice circuits cost approximately $2639 per month. Disregarding

Table 1.6 Cost comparisons: analog vs. digital circuits.

Type of circuit	Cost per bps per month*
Analog voice-grade circuit	
19.2 kbps	0.00449
Digital circuits	
2.4 kbps	0.0653
4.8 kbps	0.0326
9.6 kbps	0.0163
56.0 kbps	0.0065
T1-carrier (1.536 Mbps effective)	0.0014
T3-carrier (43.008 Mbps effective)	0.0007

*Cost computed based upon interoffice channel costs contained in Table 1.5 and equipment amortized over 36 months.

the cost of multiplexing equipment, approximately 15 (2639/182) analog circuits would justify the use of a T1-carrier facility that could support 24 voice circuits.

Another method of comparing analog and digital transmission facilities is on an aggregate transmission capacity basis. The highest data transmission rate obtainable on an analog circuit using commonly available modems is 33.6 kbps. When a 33.6 kbps data rate is compared to the 1.544 Mbps data transmission rate of a T-carrier facility, that facility can carry the equivalent of approximately 46 analog data channels even though its cost may be less than one-third that number of analog circuits.

Considering the effect of compression

The preceding economic analysis indicates that for most networking situations the use of digital transmission facilities on a per bps basis is more economical than the use of an analog transmission facility. However, the analysis did not consider the effect of data compression which is very commonly employed with modems and which is also included as a feature in some DSUs and provided by third party vendors for use on T1-carrier transmission facilities in the form of a stand-alone compression unit. Thus, let's turn our attention to the effect of compression and how it alters the cost comparison between the use of analog and certain types of digital transmission facilities.

Most modern modems support the ITU V.42 method of data compression. Although the actual effect of compression varies with the susceptibility of data to being compressed, you can normally expect to achieve an average compression ratio of 2.5:1. This means that on the average for every 2.5 bits transmitted, one bit will actually flow between modems. Viewed another way, this means that a 33.6 kbps modem with data compression enabled on the average can be expected to transfer data at 33.6 kbps * 2.5, or 84 kbps. Using that throughput, the cost of a 500-mile analog circuit operating at 33.6 kbps on a per bps per month basis, including the effect of data compression, becomes

$$\frac{(2 * \$200/36 \text{ months}) + 0.28/\text{mile} * 500 \text{ miles}}{84 \text{ kbps}}$$

$$= 0.0018 \text{ per bps per month}$$

Although data compression performing DSUs can be obtained for use on 56 kbps digital circuits, they are not presently available for use with other digital transmission facilities. Separate stand-alone compression performing devices were available for use on a T1 circuit when this book was written; however, I could not locate any products for use on low speed digital circuits. Thus, we will conclude this section by examining the economic effect of compression on the use of 56 kbps and T1 circuits.

The cost of 56 kbps compression performing DSUs was approximately $750 when this book was written. Similar to a compression performing modem, you can expect a compression performing DSU to provide an average compression ratio of 2.5:1. Thus, the effective throughput of a compression performing DSU becomes 56 kbps * 2.5, or 140 kbps. Then the monthly cost on a per bps basis of a 500-mile 56 kbps digital circuit using compression performing DSUs becomes

$$\frac{(2 * \$750/36 \text{ months}) + 0.68/\text{mile} * 500 \text{ miles}}{140 \text{ kbps}}$$

$$= 0.0027 \text{ per bps per month}$$

Third party stand-alone compression units for use on a T1 transmission facility could be obtained for $2000 per device when this book was prepared. Such products also provide an average compression ratio of 2.5:1. Thus, the use of a pair of stand-alone compression devices can be expected to boost the effective transmission on a T1 circuit to 1.536 Mbps * 2.5, or 3.84 Mbps. The monthly cost on a per bps basis of a 500-mile T1 digital circuit using stand-alone compression performing devices and conventional CSUs then becomes

$$\frac{(2 * (\$850 + \$2000)/36 \text{ months}) + 4.06/\text{mile} * 500 \text{ miles}}{3.84 \text{ Mbps}}$$

$$= 0.0006 \text{ per bps per month}$$

Note that when the effect of compression is considered, the use of an analog circuit will normally be more economical than the use of most types of low speed digital transmission facilities. Thus, a common question readers may have is why more and more organizations are replacing their analog circuits with digital transmission facilities. The answer to this question is twofold—reliability and bandwidth. Concerning reliability, digital transmission facilities have a much lower error rate than analog circuits, even though almost all analog circuits are converted to digital for routing between communications carrier offices. The difference between the use of amplifiers and repeaters on the local loop between a communications carrier's office and the subscriber has a profound effect upon the difference in the error rate between analog and digital circuits. This in turn affects the retransmission of data, since most modern communications protocols correct errors by retransmission. This in turn lowers the effective throughput obtainable on an analog circuit more than on a digital circuit.

Concerning bandwidth, although analog voice grade lines are still commonly available, it is often difficult, if not impossible, to obtain what is referred to as a wideband analog transmission facility which at one time represented the only type of high speed transmission facility that could be obtained from a communications carrier. Today organizations that require a high speed communications facility to obtain Internet access, link geographically separated LANs, or support the transmission of voice, data and video on a common transmission facility, must turn to the use of different digital transmission facilities.

Although the use of digital transmission facilities is rapidly increasing, it should be noted that this is not true for all types. Certain types of low speed digital transmission facilities, such as the original series of Dataphone Digital Service (DDS) offerings that support data transmission at 2.4, 4.8 and 9.6 kbps and other relatively low data rates, can be expected to eventually achieve the status of the Dodo bird due to their relatively high cost per bit and the availability of higher speed digital transmission facilities whose cost is significantly lower.

Path to ISDN, ATM, SDH and SONET

ISDN is defined by the ITU as 'a network, evolved from the telephone network, that provides end-to-end digital connectivity to support a wide range of services, including voice and non-voice services, to which users have access by a limited set of standard multipurpose customer interfaces'. ISDN services fall into three general categories: basic rate, primary rate, and broadband.

The basic rate interface consists of two 64 kbps channels that can be used to carry voice or data and a 16 kbps channel that carries supervisory and signaling information. Each 64 kbps channel is known as a bearer channel, giving rise to the term 'B channel'. The supervisory and signaling channel is called a D channel. These three channels, which are designed to provide a subscriber interface from a private branch exchange (PBX) to the end-user, are carried on a single twisted pair wire at a composite data rate of 144 kbps and are referred to as 2B+D.

The primary rate interface is designed to provide an inter-connection between ISDN PBXs and the ISDN network. The data transmission rate of the primary rate interface is based upon the use of current T-carrier facilities. Although T- and E-carrier facilities and the ISDN primary rate allocate channels differently, both operate at a rate of 1.544 Mbps in North America and 2.048 Mbps in countries that follow the ITU standard.

When the basic frame format used on a North American T1-carrier circuit is examined, it is seen that it consists of 24 eight-bit time slots plus a framing bit, resulting in a frame containing 193 bits. The frame frequency is 8000 frames per second, resulting in each time slot having a 64 kbps capacity, as illustrated in figure 1.8A. When voice is carried in a time slot, one bit in every 48 bits may be 'robbed' for signaling, resulting in the data transmission rate being limited to 56 kbps. In Europe, the E1-carrier circuit consists of thirty 64 kbps time slots used for data and/or voice, as well as a time slot used for signaling, and a time slot used for synchronization. This is illustrated in Figure 1.8B. In North America, the ISDN primary rate interface consists of 23 B channels used to carry voice or data and a D channel that carries all signaling information as illustrated in Figure 1.8C. In Europe, a 30B+D PRI format is used for ISDN, with all signaling information moved to the D channel. Thus, a T-carrier circuit provides the stepping stone from the high speed networks of today to ISDN primary rate interface services of tomorrow as long as the equipment used can correspond to ISDN PRI framing requirements.

Recognizing the evolving requirements of organizations to transport voice, data and video on an end-to-end basis over a scalable transmission facility resulted in the development of Asynchronous Transfer Mode (ATM) technology. Transporting

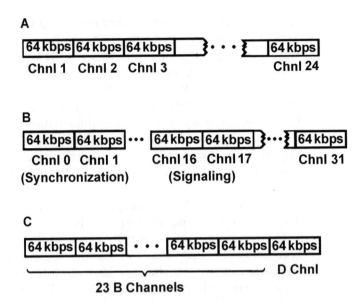

Figure 1.8 Framing comparisons: A North American T1-carrier frame format; B European E1-carrier frame format; C North American primary rate ISDN frame format

voice, data and video in 53-byte cells, this relatively new communications technology permits data to flow from LANs and PBXs in a common format over communications carrier transport facilities. ATM, like ISDN, requires the use of a digital transport mechanism. That mechanism can range from a T1 line to an optical transport facility. Concerning optical transport facilities, as we examine the use of different digital transport facilities we will also examine how T- and E-carrier circuits can be transported via different types of optical carriers (OCs), as well as the use of an optical carrier ring configuration which provides users and communication carriers with a high level of redundancy and reliability. Thus, the use of digital transmission provides a path to ISDN, ATM, SDH and SONET.

Disadvantages

Now that we have examined the advantages associated with digital signaling, let us focus our attention upon some of the problems associated with this technology.

Loss of precision

Transmitting a voice conversation on a digital network requires the conversion of analog signals to digital signals. Since an analog signal is continuous, it can take on an infinite number of values. In comparison, digital signals represent discrete values. Thus, the conversion from analog to digital can result in a loss of precision. This loss, known as quantizing noise, can result in a distortion of a voice conversation when the digital signals are reconverted into their analog form. In Chapter 2, you will find additional information concerning quantizing noise.

Analog system interface

Prior to the use of digital signaling, ringing, on-hook and off-hook status, as well as dialed digit information was conveyed by the use of relatively high current and high voltage. Digital circuitry cannot tolerate high current or voltage, resulting in the necessity of adding protective circuitry to digital devices. Although protective circuitry is most effective in eliminating most high current and voltage problems, it slightly raises the cost of interfacing digital devices to existing analog facilities.

√ Increased bandwidth requirements

In 1924, Nyquist showed that the maximum signaling rate on a channel expressed in baud (B) was related to the bandwidth (W) expressed in Hz as follows:

$$B = 2W$$

The preceding limitation avoids intersymbol interference, a condition in which one transmitted signal interferes with a succeeding signal.

Although the passband of an analog telephone channel is 3 kHz, in actuality approximately 4 kHz are used due to the bending of the attenuation-frequency response at the cutoff filter frequencies and the separation of one channel from another when frequency division multiplexing is used by a communications carrier to combine several analog signals onto a trunk routed between carrier offices. Thus, the support of 24 channels in an analog carrier system would require 4 kHz × 24 or 96 kHz.

The signal rate of 1.544 Mbps for a North American T1-carrier requires a bandwidth of 772 kHz, based upon the Nyquist relationship between signal rate and bandwidth. Thus, a T1-carrier supporting 24 voice channels requires eight times the bandwidth of an analog carrier system.

Need for precise timing

Digital signaling techniques employ synchronous transmission. Since synchronous systems require the detection of the presence of bits by sampling at precise times, timing or clocking information is required. Not only does this raise the cost of digital signaling, but, in addition, it results in a variety of errors when timing information becomes distorted or lost.

Cost

As noted from the previously performed economic comparisons, certain digital transmission facilities may be more costly than analog facilities, especially when the effect of data compression is considered. In general, low-speed digital transmission facilities will cost more than analog circuits used for data transmission, while a T1- and E1-carrier circuit capable of supporting 24 or 30 voice channels can be economically justified by a requirement for less than half that number of channels.

1.4 BANDWIDTH UTILIZATION

In concluding this chapter we will examine a few modern applications and their effect upon bandwidth utilization. Doing so will further illustrate the rationale for the growth in high speed digital transmission.

Internet access

One of the most popular applications during the late 1990s for individuals and businesses is Internet access. From the business perspective, Internet access can be represented as providing employees with the ability to access the Internet, or customers and employees with the ability to access organizational computational facilities such as Web servers connected to the Internet. To obtain an appreciation for the rationale for high speed digital networking, let's turn our attention to the display of a Web page at a popular online auction site.

Figure 1.9 illustrates the home page of Onsale viewed through the use of a Netscape browser. This page is typical of many Web pages in that it consists of a mixture of text and graphic images,

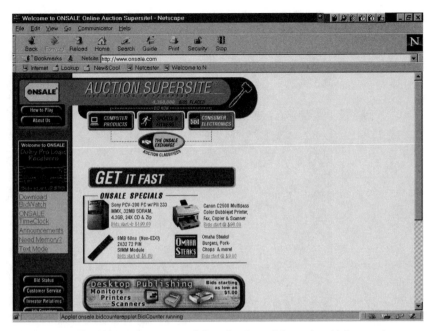

Figure 1.9 The Web home page of Onsale viewed through a Netscape browser. The use of graphics considerably increases the quantity of data that must be transmitted

with many graphics in the form of maps which, when clicked upon, route you to a predefined location on a server.

You can use Netscape or another browser to examine the source coding used to construct a Web page. Doing so lets you examine the type of images used and whether or not they are associated with a map. Figure 1.10 illustrates a portion of the source HyperText Markup Language (HTML) used to create the display shown in Figure 1.9. In examining Figure 1.10 you will note several HTML statements that reference GIF files. Although the source code does not indicate the size of graphic files, you can easily determine their size by saving them to disk. To do so you would move your cursor over an image and click on the right mouse button which displays a menu. That menu includes an option for saving the selected image.

Until the advent of the World Wide Web most communications were in the form of interactive text query–response. Instead of a display similar to Figure 1.9 which contains a large number of graphic images, most computer displays would show up to 80 columns by 25 rows of characters, or a maximum of 2000 characters. Today most computer applications result in the display of a number of images as well as text. To illustrate the effect of images, I saved the images shown in Figure 1.9 to ascertain their

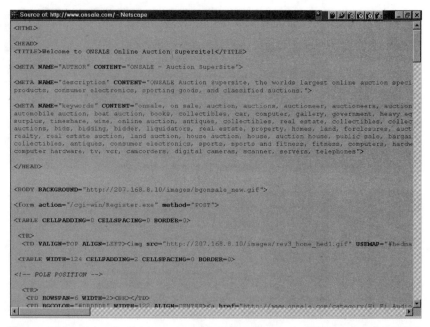

Figure 1.10 Using the 'view source' Netscape browser option to view the HTML code used to create the Onsale home page. Viewing the HTML code enables the types of graphic images to be noted

size. Figure 1.11 shows the results of a directory listing indicating the size of nine images from Figure 1.9 that were saved. Note that the cumulative size of those files is approximately 68 000 bytes, which represents an increase by a factor of 34 (68 000/2000) over the quantity of data displayed on a text-only screen. Thus, the ability of employees to access other Web sites as well as to access your organization's Web site in a reasonable time is highly dependent upon the transmission capacity of your organization's Internet connection. Today just about every Web site is connected to an Internet Service Provider via a digital transmission facility, and the growth in Web access, with traffic doubling every 100 days, is a driving force for organizations installing higher speed digital transmission facilities.

Electronic mail

Until the late 1980s and early 1990s most electronic mail messages consisted of 500–1000 characters. Since then it has become quite common for persons to attach program listings, voice recordings and graphic images to their electronic mail messages. Figure 1.12

Figure 1.11 Examining the data storage and transmission requirements of nine images located on the Onsale Web home page

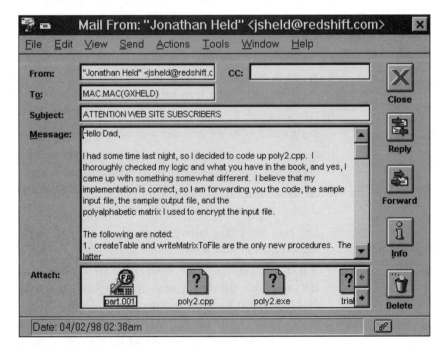

Figure 1.12 An electronic mail message containing several attached files, including C++ source and executable files

illustrates an example of an email message from this author's son. Instead of 500–1000 characters commonly transmitted as an electronic mail message, this email, including attached C++ executable files, required the transmission of almost 600 000 bytes of data. Although the total quantity of data transmitted with the electronic mail message illustrated in Figure 1.12 more than likely represents an exception to common messages, it illustrates the fact that file attachments are a reality. Today many businesses commonly email a variety of documents between organizational locations as file attachments. When coupled with the evolving use of networks originally developed for data transmission to also transport digitized voice conversations, the requirements for bandwidth are literally exploding. This in turn can be considered as the engine pulling the demand for high speed digital networking.

REVIEW QUESTIONS

1 Why is it possible for an amplifier to create an error condition?

2 How can an automatic and adaptive equalizer cause a protocol timeout?

3 What is the primary advantage associated with the use of repeaters instead of amplifiers?

4 What is the purpose of a guard band, and why does it decrease the efficiency of a frequency division multiplexer?

5 Discuss five key advantages of a digital transmission system in comparison to an analog transmission system.

6 Can you assume that digital transmission facilities are more economical than analog transmission facilities? Why?

7 Discuss the effect of compression upon the cost of analog and digital transmission facilities.

8 Explain why the conversion of an analog signal into a digital signal can result in a loss of precision.

9 What would be the difference in bandwidth requirements between 32 analog voice channels and an E-carrier digital transmission facility operating at 2.048 Mbps?

10 What are three of the driving forces which consume large amounts of network bandwidth, resulting in many organizations upgrading to high speed digital transmission facilities?

2

FUNDAMENTALS OF DIGITAL SIGNALING, CODING AND CLOCKING

In this chapter, we will examine three interrelated topics to obtain a firm understanding of the characteristics and operation of digital networks. To understand why digital signals are modified prior to transmission on a digital facility, we will first examine the characteristics of several types of digital signals. This examination will provide you with an understanding of the selection of the signaling techniques used on communications carrier digital facilities and why user equipment must be compatible with those techniques.

A comparison to signaling, which in many instances is considered synonymous, is the encoding method used to convey information over a digital transmission facility. In this book, to distinguish the two, we will consider signaling to represent the manner by which each bit is represented in a digital format on a circuit. In comparison, we will consider encoding to represent the manner by which a sequence of n bits are first converted to m bits prior to their placement on a circuit via a signaling method, with $n > 1$ and $n \neq m$. In examining encoding we will focus our attention on one method used on a fiber optic-based LAN, since that local area network is being used as a transport facility into the communications carrier digital transmission hierarchy.

In concluding this chapter we will examine clocking, timing and synchronization. Since digital networking connectivity is highly dependent upon the availability of a reliable timing reference, we will examine several methods that can be used to synchronize equipment to include the hierarchical clock system employed by digital network operators.

2.1 DIGITAL SIGNALING METHODS

One of the most critical issues to be addressed in the design of digital transmission facilities is the method by which binary data will be encoded as signal elements for transmission. The selection of one signaling method over another affects both the cost of constructing transmission facilities and the resulting quality of transmission obtained from the use of such facilities. To obtain an understanding of the advantages and disadvantages of different types of digital signaling methods, let use examine the characteristics of seven types of digital encoding techniques.

Unipolar non-return to zero

Unipolar non-return to zero is a simple type of digital signaling which was originally used for early telegraphy. Today, unipolar non-return to zero signaling is used with private line teleprinter systems, as well as the signal pattern used by the RS-232/V.24 interface.

In this signal scheme, a dc current or voltage represents a mark or binary 1, while the absence of current or voltage represents a space or binary 0. This signal encoding technique, which is illustrated in Figure 2.1, is called non-return to zero because the current or voltage does not return to zero between adjacent 1 bits. When used with a transmission system, line sampling determines the presence or absence of current or voltage, which is translated into an equivalent mark or space.

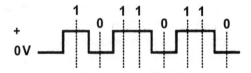

Figure 2.1 Unipolar non-return to zero signaling. In unipolar non-return to zero signaling, the current or voltage does not return to a zero value between adjacent '1' bits

That is, a receiver must have a built-in clocking mechanism which tells it when to sample the line. Typically, sampling occurs 4, 8 or 16 times per bit duration as a mechanism to enable the receiver to distinguish one bit period from another, since there could be a string of 0s or 1s. Figure 2.2 illustrates the use of a clocking source for a receiver to perform line sampling, enabling the receiver to distinguish the setting of one bit period from another.

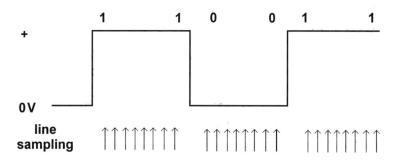

Figure 2.2 Through the use of a clocking source a receiver periodically samples the height of the received signal, enabling a '0' bit to be distinguished from a '1' bit as well as repetitions of 0s and 1s to be determined

Although it is possible to use unipolar non-return to zero signaling in low-speed transmission systems, the high data rates of most digital networks make this encoding technique undesirable. Most of the unsuitability of unipolar non-return to zero signaling is due to this signal method having a residual dc component. The residual dc component causes a direct physical attachment of transmission components in network construction. In comparison, a signal encoding method that has no dc component permits coupling to occur via the use of transformers which provide electrical isolation as well as reduce interference.

When the first high speed digital networks were developed, communications carriers looked for methods to install transmission facilities in an economical manner. One method which would result in an economical infrastructure was to employ a signaling method that would allow both power and signaling to be transported on a common wire. This would enable a common wire to deliver voltage to repeaters as well as the signal to the subscriber. However, to separate the signal from the voltage at the subscriber's location required the use of transformer coupling as illustrated in Figure 2.3. The use of transformer coupling required the use of a signaling method that did not have a residual DC component, in effect eliminating unipolar non-return to zero signaling from further consideration. In addition, as previously discussed, clocking is required at the receiver to distinguish the value of one received bit from another when a sequence of no voltage or high voltage was received. Since clocking circuitry adds to the cost of equipment, this represented a second reason why communications carriers looked for a different signaling method for use on high speed digital transmission facilities.

Transformer

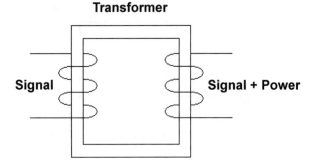

Figure 2.3 Transformer coupling which permits the signal to be removed from the signal plus power source requires a signaling method that has no DC component

Unipolar return to zero

Unipolar return to zero is a variation of unipolar non-return to zero signaling. Here the signal always returns to zero after every 1 bit. While this signal is easier to sample, as each mark has a pulse rise, it requires more circuitry to implement and is not commonly used. In addition, similar to unipolar non-return to zero signaling, unipolar return to zero results in dc voltage buildup, which means that it is unsuitable for transformer coupling and cannot be used when it is desired to transmit a signal and power on a common wire pair. Figure 2.4 illustrates an example of unipolar return to zero signaling.

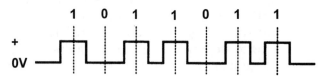

Figure 2.4 Unipolar return to zero signaling. In unipolar return to zero signaling, the signal always returns to a zero level after each '1' bit

Non-return to zero inverted

A variation of unipolar non-return to zero signaling is unipolar non-return to zero inverted (NRZI). Under the NRZI signaling method a transition at the beginning of the bit time denotes a binary '1' for the bit time, with a lack of a transition used to denote a binary '0'. The transition can be either a high to low voltage or a low to high voltage as illustrated in Figure 2.5.

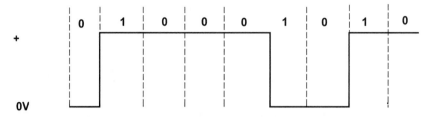

Figure 2.5 Unipolar non-return to zero inverted signaling uses a low to high or high to low transition at the beginning of the bit time to denote a binary '1', with a lack of a transition denoting a binary '0'

Similar to NRZ signaling, NRZI includes the presence of a DC component which precludes its use for transformer coupling. In addition, it does not include a synchronization capability and receivers require a separate clocking source to perform line sampling. In spite of those problems NRZI has several benefits and represents an example of differential encoding, a signaling technique used on many local area networks. As a differential encoding technique the value of each signal element is determined by comparing the polarity of adjacent signal elements. In comparison, under non-differential signaling methods such as NRZ signaling the signal is decoded by determining the absolute value of a signal element. A key benefit of differential encoding is the fact that this signaling technique permits transitions in the presence of noise to be better detected than transitions occurring when an absolute value is simply compared to a threshold.

Polar non-return to zero

In polar non-return to zero signaling a positive current is used to represent a mark and a negative current is used to represent a space. This signaling technique eliminates some of the residual dc buildup associated with unipolar signaling, since over a long period of time the number of binary 0s and binary 1s will be equal. However, since there is the possibility of having long sequences of positive or negative voltage at any point in time, you can have a residual DC component buildup that makes this signaling technique unsuitable for transformer coupling. As no transition occurs between two consecutive bits of the same value, the signal must be sampled to determine the value of each received bit. Figure 2.6 illustrates an example of polar non-return to zero sampling.

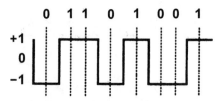

Figure 2.6 Polar non-return to zero signaling. In polar non-return to zero signaling, positive and negative currents are used to represent binary 1s and 0s, respectively. Since no transition occurs between successive 1s and 0s, line sampling must be used to determine the value of each bit

Polar return to zero

The polar return to zero signal is similar to the polar non-return to zero signal in that it uses opposite polarities of current. However, this signal returns to zero after each bit is transmitted. Figure 2.7 illustrates an example of polar return to zero signaling. Note that a sequence of 0s or a sequence of 1s can result in a dc voltage buildup. In addition, since there is a pulse that has a discrete value for each bit, sampling of the signal is not required, which reduces the circuitry required to determine whether a mark or space has occurred. Although this signaling technique did not require a clocking source, its inability to support transformer coupling precluded its use when the original digital network signaling infrastucture was developed.

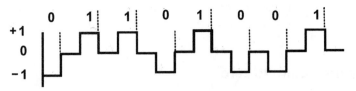

Figure 2.7 Polar return to zero signaling. In polar return to zero signaling, opposite polarities of current are used to represent binary 1s and 0s, with a return to zero current level after each bit. By returning the current value to zero after each bit, line sampling is not required to determine each bit value

Bipolar non-return to zero

Two signals similar to the return to zero encoding method, but which eliminate the problem of dc voltage component buildup, are bipolar non-return to zero and bipolar return to zero signaling. In bipolar non-return to zero signaling alternating polarity pulses are used to represent marks, while a zero pulse is used to represent a space. Figure 2.8 illustrates an example of bipolar non-return to zero signaling. Note that this encoding method does not require line sampling since the voltage levels can be examined to determine the state of the signal.

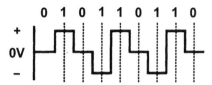

Figure 2.8 Bipolar non-return to zero signaling. In bipolar non-return to zero signaling, binary 1s are represented by alternating voltage polarities, while binary 0s are represented by a zero voltage level

Bipolar return to zero

In bipolar return to zero signaling, the bipolar signal returns to zero after each mark, as illustrated in Figure 2.9. This type of signaling insures that there is no dc voltage buildup on the line, which results in ac coupling being accomplished by the use of transformers to provide electrical isolation and reduce the possiblity of interference occurring when power and data are carried on a common line. In addition, this method of signal encoding permits repeaters to be placed relatively far apart in comparison to other signaling techniques. This signaling technique is employed in modified form on digital networks due to the economics associated with spacing repeaters further apart from one another.

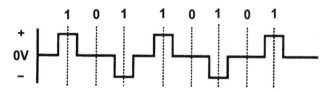

Figure 2.9 Bipolar return to zero signaling. In bipolar return to zero signaling, alternate polarities are used to represent binary 1s with the voltage returning to zero after each '1'

Fifty percent bipolar waveshape

To assist in eliminating high-frequency components that can interfere with other transmissions, digital transmission services utilize 50% bipolar waveshape or duty cycle. This signaling technique, which is illustrated in Figure 2.10, concentrates the transmitted power in the middle of the transmission bandwidth. Since the transmission quality of a channel is worse near the edges of the channel, this encoding technique minimizes the distortion that may occur to the resulting signal. This also means that the signal is well defined and any variation can be detected by equipment. The resulting bipolar pulse is also known as alternate mark inversion or AMI signaling. It is important to note that, in addition to representing a signaling method, the term AMI is also used to represent a line coding mechanism. As we will note later in this book, repeaters require a minimum number of '1' bits and cannot tolerate 15 or more consecutive zeros. To provide a periodic '1' bit AMI line coding stuffs a binary '1' into the eighth bit position of every byte transmitted which, while ensuring repeaters obtain required timing, reduces the data throughput capacity of a digital transmission facility. A second technique referred to as Binary Eight Zero Substitution (B8ZS) does not rob any bit position yet maintains a 1s density. Both AMI line coding and B8ZS, as well as its European equivalent, referred to as HDB3, are used on T1 and E1 copper circuits, with other types of zero suppression used when higher speed transmission occurs over copper circuits. Later in this book we will examine in detail different methods used to maintain a 1s density on copper circuits.

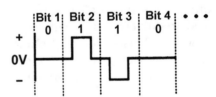

Figure 2.10 Bipolar Alternate Mark Inversion (AMI) RTZ 50% duty cycle. By concentrating the transmitted power in the middle of the transmission bandwidth, AMI signaling minimizes the distortion that may occur to a signal while eliminating dc voltage buildup on the line

Bipolar violations

Bipolar transmission requires that each data pulse representing a logical '1' is transmitted with alternating polarity. A violation of this rule is defined as two successive pulses that have the same polarity and are separated by a zero level.

A biopolar violation indicates that a bit is missing or miscoded. Some bipolar violations are intentional and are included to replace a long string of zeros that could cause a loss of timing and receiver synchronization or to transmit control information. Figure 2.11

illustrates one example of a bipolar violation. Figure 2.11A shows the correct encoding of the bit sequence 010101010 using bipolar return to zero signaling. In Figure 2.11B, third '1' bit is encoded as a negative pulse and represents a violation of the bipolar return to zero signaling technique where 1s are alternately encoded as positive and negative voltages for defined periods. In Chapter 4, we will examine several methods used to develop bipolar violations that are used to maintain synchronization when a string of consecutive 0s is encountered. These methods are commonly referred to as zero suppression codes.

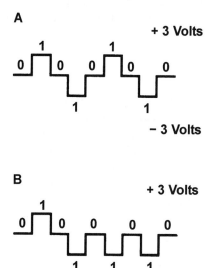

Figure 2.11 Bipolar violations. Two successive negative or positive pulses represent a bipolar violation of a bipolar return to zero signaling technique: A bipolar coding of data; B bipolar violation

LAN signaling

No discussion of digital signaling methods would be complete without examining several signaling methods used on local area networks. Although this book is primarily focused upon the use of digital transmission facilities to interconnect geographically separated locations, you should note that access to the communications carrier digital infrastructure can be obtained by the use of an FDDI ring. Thus, by examining LAN signaling we will also obtain information covering a method of accessing the communications carrier digital infrastructure that is not only growing in popularity but, in addition, offers several advantages over other access methods.

When local area networks were developed designers looked for a signal mechanism that would include a synchronization capability within the signal. A set of coding techniques referred to as bi-phase were developed which eliminated many of the limitations associated with NRZ signaling methods. Two popular bi-phase signaling techniques are Manchester and Differential Manchester which are used on Ethernet and Token Ring LANs. Since a review of these signaling methods will assist us in understanding the selection of the signaling and encoding methods for FDDI, let's first turn our attention to those two LAN signaling techniques.

Manchester

Under Manchester signaling a transition occurs at the middle of each bit period. A transition from high to low voltage in the middle of a bit period is used to denote a binary '0', while a transition from a low to high voltage in the middle of a bit period is used to denote a binary '1'. Figure 2.12 illustrates an example of Manchester signaling. In examining Figure 2.12 note that the mid-bit period transitions provide a clocking mechanism within the data stream.

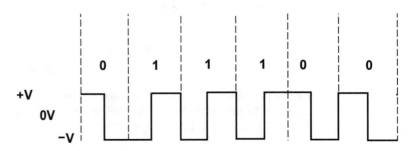

Figure 2.12   Under Manchester signaling a high to low transition denotes a binary '1' while a low to high transition denotes a binary '0'

Differential Manchester

The difference between Manchester and Differential Manchester signaling results from the manner by which binary 1s are represented. Under Differential Manchester signaling the direction of the signal's voltage transition changes whenever a binary '1' is transmitted, but remains the same for a binary '0'. Figure 2.13 illustrates an example of Differential Manchester signaling.

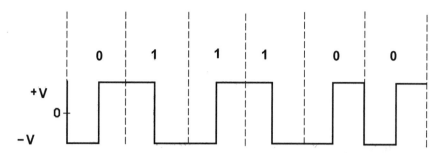

Figure 2.13 Under Differential Manchester signaling a mid-bit transition provides clocking, binary 0s are represented by the presence of a transition, and the direction of the signal's voltage transition changes whenever a binary 1 is transmitted

In examining Manchester and Differential Manchester signaling you will note that pulses have an equal positive and negative voltage which results in an absence of a dc voltage buildup. Thus, in addition to obtaining the benefits associated with a lack of a dc voltage buildup, those signaling methods facilitate error detection since the absence of an expected transition can be used to detect errors. In addition, since a predictable transmission occurs during each bit time, they permit a receiver to synchronize on the transitions, in effect representing a built-in clocking signal.

One of the problems associated with the use of a bi-phase signaling technique is the fact that they require at least one transition per bit time and can have up to two transitions per bit time. This means that the signaling rate can be up to twice the bit rate. While this fact may not appear significant, it had a major effect upon the design of signal and encoding methods used for high speed LANs to include FDDI. Thus, let's turn our attention to the topic of encoding which will enable us to examine why conventional Ethernet and Token Ring signaling methods proved unsuitable for use by FDDI.

2.2 ENCODING TECHNIQUES

As previously noted, the use of a bi-phase signaling technique can result in the signaling rate in baud being up to twice the bit rate. This meant that the development of FDDI operating at 100 Mbps would require a signaling rate of 200 MHz if a bi-phase code was used. At the time FDDI was being developed, the implementation of a 200 MHz signaling rate was pushing the state of the art and would be very expensive to implement. To reduce the required signaling rate, LAN designers turned to the use of different bit transformation schemes that were utilized in conjunction with different signaling methods. One such scheme was the use of a 4B/5B code which is specified for use with the 100 Mbps Fiber Distributed Data Interface (FDDI) optical-fiber ring topology-based LAN. Thus, let's turn our attention to the 4B/5B code.

4B/5B coding

Under the 4B/5B coding scheme four bits are encoded at a time into a five-bit code. Similar to conventional copper-based transmission, each bit in the resulting five-bit code is transmitted serially onto a fiber, with light used instead of electricity. Instead of a voltage used to denote the occurrence of a binary '1' and lack of a voltage to indicate a binary '0', each signal element is denoted by the presence or absence of a light pulse. By packing four data bits into a five-bit code, the efficiency of the coding technique becomes 80%. This means that to obtain a 100 Mbps data rate, FDDI must use a 125 MHz signaling rate. That signaling rate is substantially less than a 200 MHz rate that would be required if a bi-phase signaling scheme such as Manchester or Differential Manchester signaling were used.

Since coding by itself does not provide synchronization, a signaling method that provides synchronization is employed in conjunction with the 4B/5B coding technique. The signaling method used on FDDI optical LANs is NRZI which represents a differential encoding method. Thus, the use of 4B/5B coding and the transmission of each serial bit using NRZI signaling permits a 125 MHz signaling rate that provides synchronization through the transitions in the code.

Table 2.1 illustrates the 4B/5B code used for FDDI. Although the use of a 4B/5B code results in 32 five-bit codes, only 16 are used to encode all of the possible four-bit sequences of data. The remaining codes are assigned special meanings to indicate the beginning and ending of a data stream, the state of the transmission line, and various control indicators. In examining the data symbol entries in Table 2.1 note that they do not represent a simple placement of four bit values into five bit positions. Instead, the five-bit codes selected to represent the 16 four-bit data groups include at least two transitions for each five-bit code pattern that will be transmitted on the fiber. Since NRZI signaling is used, this ensures that there will be at least two 1s in each five-bit code since under NRZI signaling a 1 is encoded by a transition.

Table 2.1 The FDDI 4B/5B code.

General function	Code group	Assignment	
Line state symbols			
	00000	QUIET	
	11111	IDLE	
	00100	HALT	
Starting Delimiter (SD)			
	11000	1st of SD sequential pair	
	10001	2nd of SD sequential pair	
Ending Delimiter (ED)			
	01101	Terminates data stream	
Data symbols			
		HEX	**BINARY**
	11110	0	0000
	01001	1	0001
	10100	2	0010
	10101	3	0011
	01010	4	0100
	01011	5	0101
	01110	6	0110
	01111	7	0111
	10010	8	1000
	10011	9	1001
	10110	A	1010
	10111	B	1011
	11010	C	1100
	11011	D	1101
	11100	E	1110
	11101	F	1111
Control indicators			
	00111	Logical Zero (Reset)	
	11001	Logical One (Set)	
Invalid code assignments*			
	00001*		
	00010*		
	00011		
	00101		
	00110		
	01000*		
	01100		
	10000*		

These code patterns should not be transmitted as either they do not have at least two 1s in each code group or they violate duty cycle requirements. The codes marked by an asterisk (*) will be interpreted as a HALT when received.

Although the use of 4B/5B coding and NRZI signaling provides an effective and reliable transmission scheme for FDDI on optical fiber, it is not well suited for use on copper wire. The reason for this is the fact that this scheme results in the concentration of signal energy which, when employed on copper, results in a degree of radiated emissions. Those emissions would adversely effect transmission. Recognizing this problem, an encoding scheme referred to as MLT-3 is used on the twisted-pair version of FDDI referred to as CDDI where 'C' denotes copper, as well as for a version of 100BASE-T. Since this book is focused upon the digital carrier infrastructure and access to that infrastructure, with the latter obtainable via the optical version of FDDI, we will not discuss MLT-3 coding. However, when appropriate, we will discuss other coding and signaling methods later in this book.

2.3 CLOCKING, TIMING AND SYNCHRONIZATION

In a digital networking environment it is extremely important that each device attached to the network knows when to read a voltage or check for the presence or absence of a light source. If the process occurs too early or too late the incoming signal could be misinterpreted, resulting in an erroneous interpretation of the data stream. The process of ensuring that networking devices read a voltage or sample for a light source at an appropriate point in time is referred to as synchronization. Synchronization is controlled by a clocking or timing source. Thus, in this section we will turn our attention to the related topics of clocking, timing and synchronization.

Why synchronization matters

To obtain an appreciation for the reason why equipment connected to digital transmission facilities depends upon synchronization, let's examine the effect of synchronization loss.

Figure 2.14 illustrates a digital pulse sampled at three different times — t_1, t_2 and t_3. If your equipment samples the line at t_1 or t_3 it would record the pulse value as equivalent to a binary '0' instead of its actual representation of a binary '1'. This illustration presents a simplistic view of clocking as line sampling normally occurs at eight or 16 times the bit rate. However, it provides a valid illustration of the rationale for sampling a line at an appropriate point in time. If the difference between samples is less pronounced than that illustrated, a divergence between equipment clocks builds up. Eventually, at some point in time, the effect of two different clock rates, transmitter and receiver results in the widening of timing discrepancies until sampling becomes erroneous. Thus, it is extremely important that network devices are as closely synchronized as possible.

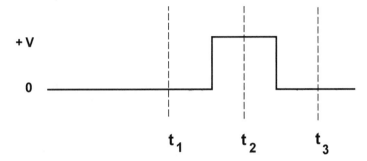

Figure 2.14 Synchronization through a clocking or timing source enables equipment connected to a digital network to sample the line at a correct point in time

The evolution of clocking

Digital transmission systems date to the 1960s when the T1 system was introduced as a mechanism to relieve cable congestion in urban areas. T1 terminal devices were designed to multiplex 24 64-kbps digitized voice channels, with each channel transporting 8000 8-bit bytes per second and each byte representing one analog sample. The actual multiplexing resulted in 24 8-bit bytes plus a framing bit used for synchronization being grouped into a frame for transmission. Thus, each T1 frame consists of 193 bits, and the 8000 frames per second transmission rate results in a line operating rate of 1.544 Mbps.

When T1 systems were initially deployed they were asynchronous, with each pair of end terminals operating at their own clock rate. The transmit and receive sides of a T1 trunk were independent of one another, enabling each device to use an internal clocking source for transmission and the received AMI coded signal as a self-derived timing/clocking mechanism. As T1 systems evolved and digital channel banks were introduced, one end terminal on the T1 circuit was designated as a master timing source and provided its own clocking. At the opposite end of the T1 circuit the other end terminal was designated as a slave and derived timing for its transmission from received data. Figure 2.15 illustrates this timing method which is commonly referred to as loop or derived timing.

If internal clocking sources are used at two ends on a circuit, it is extremely difficult for them to oscillate at exactly the same

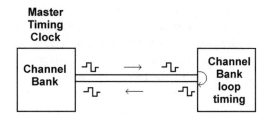

Figure 2.15 Under loop or derived timing the slave equipment uses clocking from the received signal for its transmission

frequency for a prolonged period of time. Although transmission systems are designed to tolerate a degree of clock drift, eventually the near and far end equipment will drift apart beyond the allowable tolerance. In a looped or derived timing environment where the near end depends upon the far end, a drift in the master clock would adversely affect the near end which would become cumulative as the near end used its derived clock for transmission. Both of these situations could eventually result in a situation referred to as a clock slip, where a slip is defined as a frame (193 bits) shift in time between two signals.

Since the T1 operating rate results in 8000 frames per second, the duration of each frame is 1/8000 second or 125 microseconds. When trunks transported voice, a slip had only a minor effect in the form of an audio click or popping sound being heard. When data transmission became popular, slips resulted in bit errors which caused data to be retransmitted and reduced throughput. As the use of T1 circuits increased and such circuits were routed through an evolving digital hierarchy, the occurrence of a slip at one location could spread throughout the carrier's infrastructure. To minimize this possibility, network operators developed a hierarchical clocking structure in which a master clock at the top of the hierarchy functioned as a reference for distributing clocking to equipment at communications carrier toll offices, end offices and the subscriber.

Figure 2.16 illustrates the general clocking hierarchy which evolved to support high speed digital networks. At the top of the hierarchy is a Stratum 1 clock which was originally referred to as the Bell System Reference Frequency (BSRF) when AT&T developed its digital network. This is the most accurate clock in the network and represents a Primary Reference Source (PRS) that has an accuracy of 1×10^{-11} or less as per the American National Standards Institute (ANSI) T1.101 standard. This also means that the time to the occurrence of a slip is 73 days, which results in approximately five slips per year. Although the first Bell System Stratum 1 obtained its timing from a clock whose elements decayed

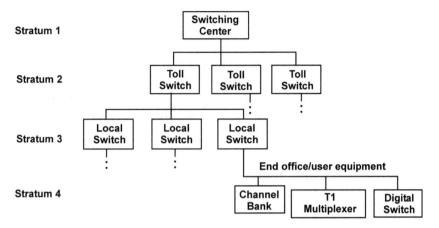

Figure 2.16   The Digital Network Clocking Hierarchy

at a fixed oscillation rate, other PRS can receive clocking from Global Position System (GPS) satellite receivers.

The second level in the hierarchy, Stratum 2, has an accuracy of 1.6×10^{-8} and delays the occurrence of the first frame slip to seven days. The Stratum 2 is used within the carrier's network at toll switching centers.

At the third level in the digital network hierarchy, Stratum 3, clocking is accurate to 4.6×10^{-6} and provides six minutes to the first frame slip. Stratum 3 is commonly used in the communications carrier hierarchy at local switches. At the bottom of the digital clocking hierarchy are channel banks and end terminals. Such equipment typically use crystal oscillators known as Stratum 4 clocks. The accuracy of a Stratum 4 clock is approximately 15 slips per minute.

Modern timing problems

The breakup of the Bell System resulted in each local operating company having to develop their own timing system to distribute clocking to their offices. Similarly, long-distance communications carriers developed their own timing systems to deliver clocking to their offices. While the installation and operation of point-to-point digital transmission facilities can occur with the use of either separate or derived or loop timing, if your circuits are configured for routing through a digital access and cross connect system (DACS), are configured as a ring, or have channels dropped and/or inserted, reliance upon equipment timing can result in the displacement of pulses known as jitter that eventually results in the loss of synchronization. Thus, it is quite common for user equipment to obtain derived timing from a carrier's Stratum clock. In doing so there are certain guidelines you should consider which we will examine in the last section in this chapter. However, instead of having to depend upon a single master clock, it is also possible to transmit high speed digital data without any central master timing clock or the use of a timing distribution network. To do so requires the use of 'near-synchronous' operations referred to as plesiochronous transmission.

Plesiochronous operations

One of the problems associated with the use of a master clock is the fact that the breakup of AT&T and other traditional telephone companies and the emergence of numerous long-distance communications carriers led to many 'master clock' systems. This resulted in each communications carrier having its own highly accurate master clock which made inter-communications subject to slips as phase differences between the clocks of different systems gradually widened. In addition, many large communications carriers sub-divided their network into regional areas, each region having its own master clock, which resulted in intra-region slips.

The subdivision of digital transmission facilities into regional areas either by one communications carrier or resulting from several communications carriers serving a similar area resulted in the term plesiochronous to define a near-synchronous capability. In effect, North American T-carriers and European E-carriers' digital hierarchies can be characterized as plesiochronous digital hierarchies. To minimize timing discrepancies between multiple T- and E-carriers as they are multiplexed resulted in a bit stuffing method of synchronization.

Bit stuffing

When lower level T- or E-carrier signals are multiplexed, the higher level signal is designed to run slightly faster than the composite throughput of the lower level signals. Certain bit positions in the composite signal are reserved as dummy bits to compensate for timing differences between subordinate systems. Thus, bit stuffing represents a method to facilitate the merging of signals without requiring any central master timing clock nor requiring a timing distribution network. Later in this book we will examine the composition of DS2 and DS3 signals and the use of bit stuffing to compensate for timing differences between lower level signals multiplexed onto a higher level signal.

Clocking guidelines

As a general rule you will lose a Stratum layer of accuracy for each network device that redistributes a clocking signal. Thus, your network structure and the manner by which you derive clocking will govern the synchronization accuracy of your network. To illustrate this, consider Figure 2.17 which shows the derivation of clocking from a Stratum 1 master reference clock. In this example let's first focus our attention on equipment labeled A, B and C. Here the network device labeled B is shown deriving its clock from A while C is shown deriving timing from device B. Since C is now deriving its clock from B, its clock source is now considered a Stratum 2 clock source which is derived or tied to the master clock. If we focus our attention on devices A, D, E and F in Figure 2.17, we see a similar situation. Now device D is considered to have a Stratum 2 clock while device E is considered to have a Stratum 3 clock. Turning once more to Figure 2.17, let's focus our attention upon device G. Since this end station is directly connected to a Stratum 1 clocking source, it receives a Stratum 1 signal.

When you design a private digital network it is important to construct a network timing diagram similar to the one presented in Figure 2.17. That timing diagram should indicate what equipment derives its clocking from a clock source. Similar to the operation of

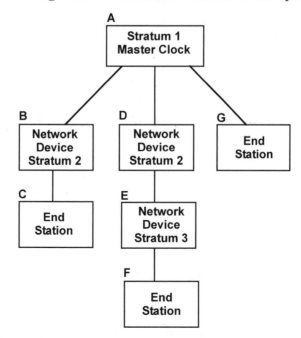

Figure 2.17 The derivation of clocking from a Stratum 1 master reference clock

Ethernet bridging, a timing diagram should not contain any closed loops. This is because a problem in one direction will automatically cause a problem to occur in the reverse direction.

While the previously presented timing information is applicable to most networking situations, the timing and synchronization of SONET and SDH optical systems represent a special synchronization situation. While the preferred method of synchronization is to use clocking from a communications carrier central office, there are several additional timing methods available for consideration. Since an understanding of those methods requires an overview of the optical transport, we will defer optical synchronization until Chapter 8 in which we cover SONET and SDH.

REVIEW QUESTIONS

1 Why was it important for communications carriers deploying early digital transmission networks to obtain a signaling method that allowed transformer coupling?

2 What is the difference between unipolar non-return to zero signaling and unipolar return to zero signaling?

3 Why is unipolar return to zero signaling easier to sample than unipolar non-return to zero signaling?

4 Why must line sampling be used to determine the value of each bit when polar non-return to zero signaling is used?

5 What are two advantages of bipolar return to zero signaling?

6 Describe the difference between Alternate Mark Inversion (AMI) used as a signaling method and as a line coding method.

7 Why is a bipolar alternate mark inversion signaling technique used on digital transmission facilities?

8 What is the difference between Manchester and Differential Manchester signaling?

9 How would the bit sequence 011001 be represented under Manchester and Differential Manchester signaling?

10 Explain why FDDI uses a 4B/5B coding scheme and NRZI signaling.

11 Describe two synchronization methods that can be used on point-to-point digital transmission facilities.

12 What is a slip?

13 Why is it important to construct a timing diagram when you design a private digital network?

3

VOICE DIGITIZATION TECHNIQUES

The earliest method of voice digitization formed the basis for the design of the T-carrier in North America and the E-carrier in Europe. This technique, which is referred to as Pulse Code Modulation (PCM), also resulted in the design principles associated with the development of a hierarchy of T- and E- communications carrier transmission systems as well as the development of SONET and SDH systems. In addition, numerous third party products to include T- and E-carrier multiplexers and Private Branch Exchanges (PBXs) were designed to support time slots formed by the use of PCM encoded voice. Today PCM is primarily used to transport calls originated over the switched telephone network. PCM was followed by several additional voice digitization techniques that were employed by communications carriers and third party equipment vendors. Such voice digitization techniques fall under the general category of waveform coding and provide a high quality reconstruction of the analog signal that was digitized. Two other categories of voice digitization include vocoding and hybrid coding. While voice digitization methods that fall into those two categories are not normally used by communications carriers and had nothing to do with the design of their infrastructure, such methods are commonly supported by third party equipment developers. Due to the potential advantages associated with methods that fall into those categories we will also examine those methods in this chapter. However, since the primary purpose of this chapter is to obtain an understanding of the association between voice digitization and the development of the communications carrier digital infrastructure, we will focus our attention primarily upon waveform coding techniques in this chapter.

3.1 WAVEFORM CODING

Waveform coding represents a category of voice digitization whereby samples of a signal are taken on a periodic basis, with each sample converted into a digital value. Examples of voice digitization methods that fall into the category of waveform coding include pulse code modulation (PCM), adaptive differential pulse code modulation (ADPCM), continuous variable slope delta modulation (CVSD) and variable quantizing level (VQL).

Pulse code modulation (PCM)

The earliest method used by communications carriers to convert the continuously varying subscriber voice (analog) signal into a digital data stream was PCM. PCM is still the most common technique utilized to digitize voice due to the prior investment of communications carriers in that technology. This technique requires three processing steps—sampling, quantization and coding.

Sampling

During the PCM sampling step, the analog signal is sampled 8000 times per second. This sampling rate is based upon the Nyquist theorem which states that to truly represent a signal the sampling rate must be twice the highest frequency of the signal to be transmitted.

Although the passband of a voice channel created by the use of low pass and high pass filters is approximately 3000 Hz, the filters do not instantaneously cut off frequencies below 300 Hz and above 3300 Hz. At very low power levels the voice bandwidth spreads out to approach approximately 4000 Hz, resulting in the selection of a sampling rate of 8000 times per second or every 125 microseconds.

The level or amplitude of each sample is determined by a coder–decoder (codec) and a pulse proportional to the amplitude at the sampling instant is created. Figure 3.1A illustrates an analog wave that is to be digitized. The resulting signal, based upon the previously described sampling, is called a pulse amplitude modulation (PAM) wave which is illustrated in Figure 3.1B.

In examining the PAM waveform shown in Figure 3.1B, note that the PAM process results in a sequence of 8000 samples per second, each sample representing the voltage of the wave at the time the sample occurred. Also note that the height of each sample has an infinite number of possible values since the samples represent the height of the wave at defined times. Since the sequence of PAM

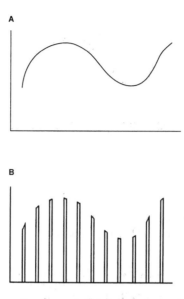

Figure 3.1 Pulse amplitude modulation (PAM). A pulse amplitude modulation (PAM) signal consists of a series of voltages which represent samples of an analog wave that were taken at predefined times: A analog wave; B pulse amplitude modulation waveform

samples represent information in an analog form they are not suitable for transmission over relatively long distance. Therefore the sequence of PAM samples need to be digitized to be transmitted on a digital carrier. However, since each sample could have an infinite number of values they must first be converted into a limited number of discrete amplitudes. This conversion process is called quantization and represents the second PCM processing step.

Quantization

As previously noted, the process of reducing a PAM signal to a limited number of discrete amplitudes is called quantization. This second part of the encoding process is necessary since the PAM samples can represent an infinite number of amplitude levels. Since the PAM waves represent a continuous height that can have any value within a range, while quantization results in a limited number of discrete amplitudes, this process will introduce errors. Such errors are called quantization noise, and represent the difference between the resulting discrete coded value of the PAM samples and the actual values of the samples. Thus, the number of discrete steps used to code PAM samples affects both the potential quantization error and the number of bits required to encode the resulting PCM data element.

Experiments have shown that the use of 2048 uniform quantizing steps provides the ability to obtain a sufficient capability to reproduce a voice signal of high quality. For 2048 quantizing steps, an 11-element code (2^{11}) would be required. Using a sampling rate of 8000 samples per second, the data rate required to digitize a voice channel would become 88 000 bps. This data rate is reduced by the third step in the PCM process—coding.

Coding

To reduce the number of quantum steps, two techniques are commonly used—non-uniform quantizing and companding prior to quantizing, followed by uniform quantizing. In non-uniform quantizing, the step assignments used for encoding are changed so that large steps are assigned to portions of high and low amplitudes, while smaller steps are assigned to intervals between those steps. In companding, the analog signal is first compressed prior to coding, followed by expansion after decoding. This permits a finer granularity in the form of additional steps to the smaller amplitude signals.

The objective of each technique is to reduce the number of quantum steps to 128 or 256, enabling either seven bits (2^7=128 quantum steps) or eight bits (2^8=256 quantum steps) to be used to encode each PAM sample.

Most PCM systems employ companding with non-uniform quantization. The compression, as well as eventual expansion, are based upon logarithmic functions that follow one of two laws: the A law and the 'mu' (μ) law.

Both North American and Japanese PCM systems employ μ law encoding, while European PCM systems utilize A-law encoding. Each technique defines the number of quantizing levels into which samples can be encoded and how those levels are arranged. Both laws use a logarithmic scale for quantization that favors the low amplitude portions of a signal where the majority of speech energy resides.

The μ-law divides the quantization scale into 255 discrete units of two different sizes called segments or chords and steps. There are 16 segments or chords, eight used to represent positive signals and eight used to represent negative signals. Figure 3.2 illustrates the general spacing of chords. Note they are spaced logarithmically, with each chord larger than the preceding chord. Within each chord there are 16 steps spaced linearly. Thus, the step size is larger in larger (higher) chords. Since the zero level is shared, there are $16 \times 16 - 1$, or 255, levels available for use. In comparison, coding using the A law results in a 13-segment approximation

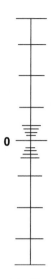

0

Figure 3.2 Chords are spaced logarithmically with each larger than the preceding

since there are six segments and the segments passing through the origin are colinear and counted as one segment.

The curve for the A law is plotted from the formulae

$$Y = \frac{AX}{1 + \log A} \qquad 0 \leqslant X \leqslant 1/A$$

$$Y = \frac{1 + \log (AX)}{1 + \log A} \qquad \frac{V}{A} \leqslant v \leqslant V$$

where V is the maximum input voltage and v is the instantaneous input voltage.

The curve for the μ law is plotted from the formula

$$Y = \frac{\log (1 + \mu X)}{\log (1 + \mu)} \qquad -1 \leqslant X \leqslant 1$$

For the above formulae

$$X = \frac{v}{V} \text{ and } Y = \frac{i}{B}$$

where i represents the number of the quantization step commencing from the center of the range while B represents the number of quantization steps on each side of the center of the range.

The relationship between the linearity of an input signal and the resulting compressed input coded value of the signal is based upon the selection of values for the parameters A and μ. The selection of values for those parameters determines the range over which the ratio of signal-to-distortion remains relatively constant. This range is known as the dynamic range and has a value of approximately 40 dB. The dynamic range value is obtained using a value of 87.6 for A for the A law. For the μ law, μ is set to a value of 100 in older systems that use a seven-segment approximation of the logarithmic curve. For more modern 16-segment approximations, μ is set to a value of 255.

Figure 3.3 illustrates a general plot of quantization in the North American D2 PCM system. This system uses the μ law with a value of 255 assigned to μ. For clarity, only the positive portion of the curve is assigned values. Note that segment 6 has been exploded to illustrate that although the segments, which are also known as chords, are spaced logarithmically, within each chord the steps are spaced linearly.

Table 3.1 lists the μ law PCM code table used in the North American D2 PCM system. In this system there are actually only 255 quantizing steps since step 0 and 1 use the same bit sequence to avoid a code sequence with no transitions, such as 0s only.

An examination of the entries in Table 3.1 illustrates the encoding format of PCM words. As indicated by the entries in the table, the first bit denotes whether or not the signal was above (1) or below (0) the horizontal axis. The next three bit elements identify the segment or chord, while the last four bits identify the actual step or level within the chord. Figure 3.4 illustrates the PCM word encoding format which provides several advantages over a simple assignment based upon binary bit values. First, with a conventional encoding system, equipment would have to read eight bits to tell if a

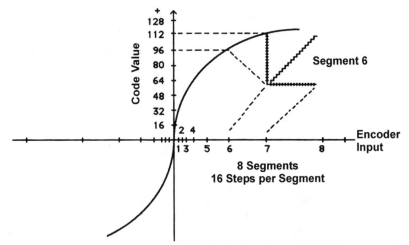

Figure 3.3 Quantizing curve used in North American D2 PCM systems. The North American D2 PCM system uses 16 segments (eight shown for clarity) spaced logarithmically based upon μ set to a value of 255

Table 3.1 North American D2 μ-law.

Code level		Digit number							
		1	2	3	4	5	6	7	8
255	(Peak positive level)	1	0	0	0	0	0	0	0
239		1	0	0	1	0	0	0	0
223		1	0	1	0	0	0	0	0
207		1	0	1	1	0	0	0	0
191		1	1	0	0	0	0	0	0
175		1	1	0	1	0	0	0	0
159		1	1	1	0	0	0	0	0
143		1	1	1	1	0	0	0	0
127	(Center levels)	1	1	1	1	1	1	1	1
126	(Nominal zero)	0	1	1	1	1	1	1	1
111		0	1	1	1	0	0	0	0
95		0	1	1	0	0	0	0	0
79		0	1	0	1	0	0	0	0
63		0	1	0	0	0	0	0	0
47		0	0	1	1	0	0	0	0
31		0	0	1	0	0	0	0	0
15		0	0	0	1	0	0	0	0
2		0	0	0	0	0	0	1	1
1		0	0	0	0	0	0	1	0
0	(Peak negative level)	0	0	0	0	0	0	1*	0

*One digit is added to ensure that the timing content of the transmitted pattern is maintained. Note that there are actually only 255 quantizing steps because steps '0' and '1' use the same bit sequence, thus avoiding a code sequence with no transitions (i.e., 0s only).

number was positive or negative. Using the encoding format illustrated in Figure 3.4 permits the polarity of the sample to be determined by the examination of one bit position. A second advantage of the PCM word encoding format illustrated in Figure 3.4 is that it prevents most line hits from significantly changing the value of the PCM word. As an example, the altering of a step bit would affect the value of only that step.

In comparison to the North American μ law, European systems quantize voice signals using a 13-segment approximation of the A law curve. For the most part, both techniques provide a similar level of quality for a reconstructed signal. In general, the A law produces a better signal-to-noise ratio at low levels, while the μ law has a lower idle channel noise level.

Companders

The term 'compander' is a contraction of the terms 'compressor' and 'expander' used in communications carrier systems to improve

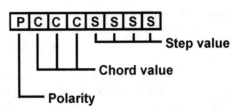

Figure 3.4 PCM word encoding format. The PCM word encoding format permits the polarity of the sample to be determined by the examination of one bit and prevents most line hits from significantly changing the value of the word

the signal-to-noise ratio of an analog signal. Through the use of companders, the dynamic range of voice signals can be compressed prior to quantization. Then, after the encoded signal is converted back into its analog form by an interpolator, an expander expands the range of the analog signal to its original range. Figure 3.5 illustrates the companding process in block form.

The compressor performs the following functions:

- It raises the power level of weak signals so they can be transmitted above the noise and crosstalk level associated with a typical communications channel.

- It attenuates very strong signals to minimize the possibility of crosstalk affecting other communications channels.

Prior to discussing the operation of a compressor–expander, a brief review of the term decibel (dB) may be in order for some readers. In the context of this book, the gain or loss of power on a circuit is expressed in decibels (dB) and is defined as

$$\text{Gain/loss} = 10 \log_{10} \left(\frac{P_O}{P_I} \right)$$

where P_O is the output or received power and P_I is the input or transmitted power. The value of $\log_{10} X$ can be expressed as 'the power by which 10 must be raised to equal X. Thus, $\log_{10} 100$ has a value of 2 while $\log_{10} 1000$ has a value of 3. Another notable property of logarithms is the fact that $\log_{10}(1/X) = -\log_{10} X$. This makes it relatively easy to compute the dB value when there is a power loss $(P_O < P_I)$.

Figure 3.6 illustrates the typical operation of a compressor and expander. The compressor accepts a 60 dB input power range, typically between 20 and 80 dB. The compressor increases the weakest sounds in power from 20 to 40 dB, while the strongest sounds are decreased in power from 80 to 70 dB.

The expander unit works exactly opposite to the compressor unit, expanding the reduced power range back to its original form. Here

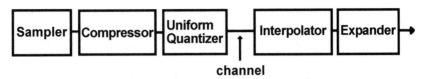

Figure 3.5 The compressor–expander (companding) process. The compressor reduces the dynamic range of a voice signal prior to quantization, while the expander expands the range of the analog signal after the interpolator restores it to an analog format

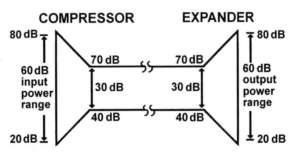

Figure 3.6 Compressor–expander operation. The compressor reduces the dynamic range of an analog signal by 30 dB, while the expander increases the signal to its original range

the shift in power is opposite to that performed by the compressor, with the weakest sounds decreased in power from 40 to 20 dB, while the strongest sounds increase in power from 70 dB up to 80 dB.

As a result of the compression quantization and encoding process, each PAM sample is encoded into either seven data bits + one signaling bit used to indicate the polarity (sign) of the sample (128 quantizing steps) or eight data bits (256 quantizing steps). For either encoding method, the resulting data rate becomes

$$8000 \text{ samples/second} \times 8 \text{ bits/sample} = 64 \text{ bps}$$

Based upon the preceding, each voice conversation digitized according to the PCM method results in a data rate of 64 kbps. Since the first level in the North American T-carrier (called T1) supports 24 digitized voice conversations, its operating rate would appear to be 64 kbps × 24, or 1.536 Mbps.

In actuality, the T1 carrier in North America operates at 1.544 Mbps. The difference, which is 8000 bits per second, is used for framing to include synchronization. For European systems, the E-carrier supports 32 channels operating at 64 kbps, resulting in a data rate of 2.048 Mbps. Since signaling and synchronization are embedded in two 64 kbps channels, the actual usable data rate of

European systems is $64\,kbps \times 30$, or 1.92 Mbps. Both North American T1-carrier framing and European E1-carrier signaling and synchronization are covered in detail in Chapter 5.

As previously mentioned at the beginning of this chapter, the selection of PCM as the first and primary voice digitization technique by communications carriers significantly affected the manner by which their transmission infrastructure was designed. The 64 kbps voice digitization rate initially became the lowest channel operating rate of digital channel banks and also resulted in third party equipment developers designing their products to operate upon data in 64 kbps increments. However, as the use of voice communications expanded, communications carriers examined the potential use of other voice digitization techniques as a mechanism to increase the number of simultaneous calls that could be supported over T- and E-carrier transmission facilities, especially on international circuits that had a limited transmission capacity and whose expansion could be extremely expensive, especially if it required the installation of an undersea cable. In addition, private organizations also wanted the ability to transmit additional calls between organizational locations interconnected by expensive T- and E-carrier leased lines used to interconnect PBXs at each location. This resulted in equipment vendors developing several additional waveform voice digitization methods which are primarily used on private networks constructed through the leasing of T- and E-carrier transmission facilities. Since the public switched telephone network is based upon the use of PCM, voice calls digitized through the use of other digitization methods are commonly reconverted via the use of a gateway for placement onto the PSTN.

Adaptive differential pulse code modulation (ADPCM)

One of the first sub-64 kbps voice digitization methods developed to enable an increase in the voice transporting capacity of T- and E-carrier transmission facilities was adaptive differential pulse code modulation (ADPCM). Although the original development of ADPCM resulted in the encoding of 64 kbps PCM into a 32 kbps data stream, subsequent versions of ADPCM support 16, 24 and 40 kbps voice digitization rates. ADPCM uses a sampling rate of 8000 samples per second, the same as used in PCM. A transcoder uses an algorithm to reduce the number of quantizing levels to 16, permitting each sample to be represented by a 4-bit word. This process, which is illustrated in the left portion of Figure 3.7, results

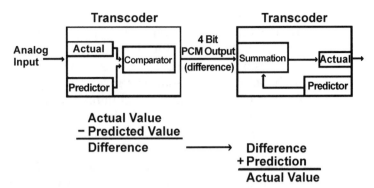

Figure 3.7 Adaptive differential pulse code modulation operation. In adaptive differential pulse code modulation (ADPCM), the predicted value is subtracted from the actual PCM value, resulting in the difference being transmitted as a 4-bit word

in the predicted value being subtracted from the actual value, permitting the difference to be transmitted in the form of a 4-bit PCM word.

The 4-bit word used by ADPCM represents the difference between the actual value of the signal and its predicted value. This difference information is sufficient to reconstruct the amplitude of the signal.

At the opposite end of the channel, another transcoder with an identical predictor performs the ADPCM process in reverse, restoring the predicted signal to the original 8-bit code. As indicated in the right portion of Figure 3.7, a summation process adds the received difference to the predicted value generated by the receiving transcoder to obtain the actual value of the signal.

The key to ADPCM is the fact that the human voice does not change significantly from one sampling interval to another, permitting a high degree of prediction accuracy. This means that the difference between the predicted and actual signal is very small and can be encoded using only four bits. In the event that successive samples should vary widely, the algorithm used for prediction will adapt to the changes by increasing the range represented by the four bits. However, as might be expected, this adaptation reduces the accuracy of voice frequency reproduction.

The first version of ADPCM resulted in the data rate of a digitized voice conversation becoming

$$8000 \text{ samples/second} \times 4 \text{ bits/sample} = 32\,000 \text{ bps}$$

Note that ADPCM results in a data rate one-half of the PCM data rate. Thus, the use of this version of ADPCM doubles the digitized voice carrying capacity of a North American T1 system to 48 channels and a European E1 system to 60 voice channels.

In actuality, the line format standard for 32 kbps ADPCM compression specifies three signaling methods. Those signaling methods permit 44 channels in a T1 along with four signaling and alarm overhead channels, 48 channels with switching occurring on 64 kbps boundaries, and 48 channels operating at 64 kbps with a bit robbed in the sixth frame for signaling. Later in this book we will examine bit robbing in detail.

Other versions of ADPCM operate at 16, 24, 32 or 40 kbps by adjusting the sampling rate while maintaining the use of a 4-bit word in place of PCM's 8-bit word. The primary use of ADPCM by communications carriers is on international circuits due to the high cost associated with establishing a long-distance transmission capability, especially when the transmission capacity is obtained by the installation of an undersea cable. Most communications carriers that employ ADPCM do so at 32 kbps; however, modern ADPCM chip-based processors support operating modes of 16, 24, 32 and 40 kbps. ADPCM operating rates other than 32 kbps are primarily used by organizations that construct their own private networks through the use of leased T- and E-carriers and the acquisition of equipment that supports different ADPCM voice digitization modes of operation.

Advantages

The major advantage of ADPCM is that its use permits the capacity of voice and voice/data systems to increase. This is because the use of an 8 kHz sampling rate and 4-bit code enables digitized voice to be reduced to 32 kbps, permitting a doubling of the number of voice conversations that can be carried on a system or lowering the data rate to support voice which allows data transmission support to increase. In addition, ADPCM requires less complex circuitry than other 8- to 4-bit encoding schemes, such as time assigned speech interpolation (TASI), continuous variable slope delta modulation (CVSD) and near instantaneous companding (NIC).

Disadvantages

Like most technology, the use of ADPCM involves several trade-offs. Until the 1980s, transcoding algorithms were not standardized. Thus for many years the system of one manufacturer had a high degree of probability that it would not be compatible with those manufactured by another vendor. Even when systems are compatible, it may not be a good idea to mix and match devices using the same ADPCM compression algorithms, since using equipment from different vendors inhibits the uniformity of diagnostic capabilities. In addition, because the patterns of voice and data differ, separate adaptive predictors must be dedicated to each, requiring a speech/data detector in each transcoder to determine which algorithm to employ.

The last major disadvantage associated with the use of ADPCM relates to its ability to transmit modulated data. In general, ADPCM provides marginal support for 4-wire modems operating above 4800 bps, since signal-to-noise level resulting from the application of a predictive algorithm is not constant.

Continuous variable slope delta modulation (CVSD)

Continuous variable slope delta modulation (CVSD) was originally used in military communications as it facilitated the encryption of analog voice conversations. Today, several T- and E-carrier multiplexer vendors market CVSD digitization modules that can be added to their equipment to digitize voice at relatively low data rates.

In the CVSD digitization technique, the analog input voltage is compared to a reference voltage. If the input is greater than the reference, a '1' is encoded, while a '0' is encoded if the input voltage is less than the reference level. This permits a 1-bit data word to represent the digitized voice signal. Figure 3.8 illustrates the resulting encoded values of a portion of an analog signal at defined sampling periods.

Early military systems sampled the analog waveform 8000 times per second, resulting in a bit rate of 8 kpbs. Although some commercial systems can also be set to that sampling rate, in general, good quality voice reproduction requires a faster sampling rate. Today, most military and commercially available CVSD systems sample the analog input at 16 000, 24 000 or 32 000 times per second, resulting in a bit rate of 16, 24 or 32 kbps to represent a digitized voice signal.

Table 3.2 lists the maximum number of voice channels North American and European T1- and E1-carrier systems can support based upon commonly used CVSD digitization rates. As might be expected the large voice channel support at low CVSD digitization rates is the major advantage obtained from this digitization method. Unfortunately, the use of CVSD modules in multiplexers is not recommended for passing modem-modulated data in a

Encoded Values

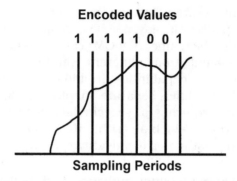

Sampling Periods

Figure 3.8 Continuous variable slope delta modulation (CVSD). Samples that have an increased height in comparison to a previous sample are encoded as a '1', while a sample that has a height less than a previous sample is encoded as a '0'

Table 3.2 T-carrier voice capacity using CVSD.

CVSD data rate (kpbs)	North America	Europe
16	96	120
24	72	90
32	48	60

digital format. Due to the problem associated with carrying modulated data, communications carriers do not use this digitization technique in their facilities.

Variable quantizing level (VQL)

Variable quantizing level (VQL) is a voice digitization technique developed by Aydin Monitor Systems to provide a 32 kbps digital stream from a telephone voice frequency (VF) channel. One of the most interesting aspects of this digitization technique is its use of four error control bits which enables the receiving end to detect and correct any single bit error that might occur in the header of a 'VQL word'. This limited measure of protection provides partial protection to the digitization's critical component, which is the header of the VQL word.

Under the VQL algorithm, the speech waveform is filtered from 3400 to 3000 Hz. To reduce the high end of the voice passband, the resulting passband is sampled 6667 times per second, and PCM encoded samples are then processed in blocks of 40, which corresponds to a 6 millisecond snapshot of the speech signal. Since there are 6667 samples per second within the 3000 Hz bandwidth, the ratio of sampling to bandwidth is approximtely 2.2. This ratio exceeds PCM's 2:1 ratio (8000/4000), which provides better fidelity of voice reproduction within the 300 to 3300 Hz frequency range.

For each block of 40 PCM encoded samples, a maximum amplitude is obtained which is then divided into 11 steps of equal magnitude. Each sample in the block is then compared to the maximum amplitude and assigned a new code which corresponds to the nearest of the 11 levels as illustrated in Figure 3.9. The data for a 6 millisecond block is then formed into a 'VQL word' which contains the maximum amplitude of the 40 samples, signaling information, forward error correction bits, and the 40 encoded samples. Due to the blocking of samples into a VQL word, an inherent delay up to 6 milliseconds is associated with this digitization technique.

In the example illustrated in Figure 3.9, assume block I in a sequence of 40 PCM samples had the maximum amplitude of all samples. Then, the height of that block, which is 44, is converted into a sequence of 11 levels. Each level is then encoded by the use of four bits plus a sign bit, resulting in each sample being represented by five bits. Next, the maximum amplitude of the sequence of 40 samples in the block, which in this instance is 44, is encoded into a header, followed by two signaling bits and four forward error correcting bits.

The four forward error correcting bits protect the maximum amplitude and two signaling bits, permitting the receiving end to detect and correct any single bit error in the 12-bit header.

Based upon the preceding, the VQL channel rate would be computed as follows:

$$\frac{6667 \text{ samples/second}}{40 \text{ samples/block}} = 166.675 \text{ blocks/second}$$

$$40 \text{ samples} \times 5 \text{ bits/samples} + 12 \text{ header bits} = 212 \text{ bits/block}$$

$$166.675 \text{ blocks/second} \times 212 \text{ bits/block} = 35\,335 \text{ bps}$$

Since this data rate exceeds 50% of the conventional PCM data rate, additional processing is used to obtain a 32 kbps rate that doubles the capacity of a T-carrier. To obtain a 32 kbps channel rate, two samples are combined to improve coding efficiency. Since each VQL sample can assume 22 possible values (11 positive and 11 negative steps), two samples taken together can assume one of 22×22, or 484, possible values. That product can be encoded in nine bits ($2^9 = 512$), which is equivalent to 4.5 bits per sample.

Using 4.5 bits per sample shortens the VQL word to

$$40 \text{ samples} \times 4.5 \text{ bits/sample} + 12\text{-bit header} = 192 \text{ bits}$$

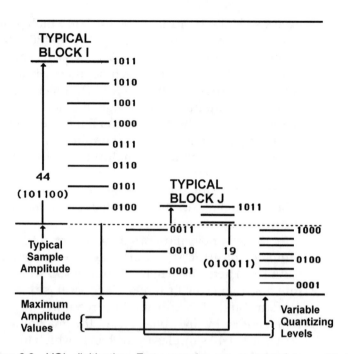

Figure 3.9 VQL digitization. Forty samples are converted to an 11-step scale defined by the maximum amplitude in the block of samples

Then

$$166.675 \, \text{blocks/second} \times 192 \, \text{bits/VQL word} = 32 \, \text{kbps}$$

which is exactly one-half of the conventional PCM data rate.

Advantages and disadvantages

Due to the encoding techniques used by VQL, as the signal amplitude decreases, the magnitude of the quantizing steps decreases almost in proportion. This results in a near-constant signal-to-noise level, which permits support of four-wire modems operating up to 9600 bps. In addition, the use of forward error correcting provides a measure of integrity to the most critical part of the encoding process, the VQL word header. Thus, VQL has a limited ability to correct itself from line hits and other disturbances.

Although VQL's performance from 300 to 3000 Hz is excellent, its design results in the loss of the high frequencies between 3000 and 3400 Hz. While the loss of high frequencies does not appreciably affect voice, it can significantly reduce the operating rate of packetized ensemble protocol modems that attempt to transmit on up to 512 carriers spaced throughout the voice channel passband. Thus, this voice encoding technique can adversely affect some modem modulation methods used to transmit data.

Another problem with VQL, as with many other voice digitization techniques, is that this method of voice encoding is not standardized.

3.2 VOCODING

The term vocoding, which represents a contraction of 'voice' and 'coding', refers to a series of voice digitization methods based upon the development of different models of human speech. Most vocoding methods are based upon the assumption that speech can be generated by exciting a linear system by a series of periodic pulses if sound is voiced, or the use of noise to represent the vocal track if sound is unvoiced. Vocoders perform an analysis of speech, extracting voice parameters that will be transmitted instead of waveform samples used by waveform coding techniques. At the receiver, the transmitted parameters are used to reconstruct voice via a synthesis process. Although vocoding methods result in a low bit rate, with a voice conversation reduced to between 2400 bps and 4800 bps, their extraction of speech parameters is performed without considering whether or not the synthesis process will generate a waveform resembling the original signal. This process carried out many times results in the reconstructed synthesized voice sounding a bit 'metallic'. In addition, the analysis performed at the transmitter many times has difficulty separating background noise from the actual voice conversation, reducing the clarity of the resulting synthesized voice reconstruction process. In spite of these problems, several types of vocoders were commonly employed on international circuits by corporations seeking to reduce the cost of their global communications bill during the 1980s. One of the most popular types of vocoding is the linear predictive vocoder that owes part of its historical evolution to a child's toy that was popular during the 1970s.

Linear predictive coding

During the 1970s a toy in the shape of an owl was introduced that represented one of the earliest commercial applications for voice synthesis. The toy, called 'Speak and Spell', included a keyboard, speaker, and red and green colored owl eyes. A child (and many adults) would press a button and the owl would randomly select a word from its memory and pronounce the word using voice synthesis. The toy operator would then use the keyboard to spell the word just pronounced. If they spelled the world correctly, the owl's green light would illuminate, otherwise the red light would illuminate. In addition to becoming a popular toy, Speak and Spell also illustrated the practicality of voice synthesis, especially when data storage was limited, which in effect is equivalent to requiring a low bit rate.

The basis behind the operation of linear predictive coding (LPC) is the analysis it performs on analog speech. Figure 3.10 illustrates the speech producing elements of the human vocal track. During LCP encoding, the analog voice input is analyzed and then converted into a set of digital parameters for transmission. At the receiver, a synthesizer recreates an analog voice output based upon the received set of digital parameters. By limiting the analysis of the voice signal to four sets of voice parameters, a very low data rate can be used to transmit voice data in digital form.

Operation

In linear predictive coding, the voice signal is first sampled by a 12-bit analog-to-digital converter. The output of the converter is then used as input to four parametric detectors. A pitch detector analyzes the data to obtain the fundamental pitch frequency at which vocal cords vibrate. Next, a voice/unvoiced detector senses whether sound is caused by the vibration of vocal cords (voice) or by sounds such as 'shhh' (unvoiced) that do not vibrate. A power detector then determines the amplitude (volume or loudness) of the

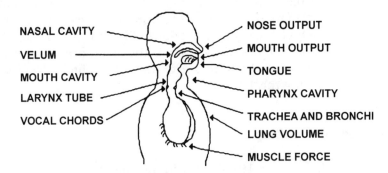

Figure 3.10 Speech producing elements of the human vocal tract. The speech producing elements of the human vocal tract are analyzed by a linear predictive encoder to synthesize a voice conversation

sound, while a spectral data decoder models the resonant cavity formed by the throat and the mouth.

Since LPC sends speech parameters rather than the amplitude of waveforms, it is actually a method of speech synthesis.

Equipment application

During the 1980s LPC was integrated into several devices collectively known as voice digitizers that can operate at 2400 or 4800 bps. Typically, data from the four LPC detectors is stored in a 54-bit buffer that is released every 22.5 milliseconds. This results in 44.444 samples per second which, when multiplexed by 54 bits per sample, permits an analog voice conversation to be digitized at 2400 bps.

Figure 3.11 illustrates two typical voice digitizer applications. In Figure 3.11A, two voice digitizers are connected to a PBX, enabling

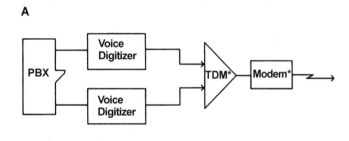

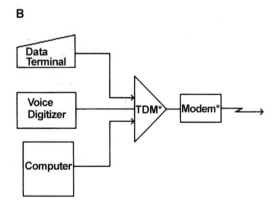

Figure 3.11 Voice digitizer applications: A multiple voice channel derivation; B mixed voice and data networks; *the time division multiplexer (TDM) and modem can be replaced by a DSU that contains a built-in synchronous time division multiplexer

the output of both devices to be multiplexed onto a common circuit. Using this technique, organizations can derive two or more voice channels from one analog circuit.

In Figure 3.11B, the integration of one voice channel into a data network is illustrated. In this example, the output of the voice digitizer is multiplexed with digital data sources onto a common data channel.

Constraints to consider

The primary constraints associated with the use of devices that employ linear predictive encoding are the cost of equipment and the fidelity of the reconstructed signal. When voice digitizers were introduced their cost commonly exceeded $7000 per unit, making their use economically feasible only for international long distance. Although voice digitizer prices have significantly fallen, most linear predictive coding methods have been superseded by hybrid coding built into multiplexers and Frame Relay Access Devices (FRADs).

With respect to the fidelity of reconstructed voice, LPC synthesizes both the speaker's conversation and any background noise. Due to the method used to synthesize voice conversations, background noise is accentuated, which can result in some disturbance to the reconstructed analog signal. Thus, the use of LPC is more suitable to conversations originating in an office environment than for conversations on a factory floor where there may be a large amount of background noise.

A second fidelity problem associated with LPC is its low bit rate. This results in a small bit error rate affecting a larger portion of speech quality than an equivalent error rate on a PCM or ADPCM system. In some situations, a low bit error rate can actually result in the reconstructed voice appearing as speaker stuttering.

3.3 HYBRID CODING

Hybrid coding represents a combination of waveform coding and vocoding which results in a high quality of reconstructed speech transmitted at relatively low bit rates. Under hybrid coding speech is passed through a vocal track predictor and a pitch predictor. The output of the prediction process, which represents synthesized speech, is used to generate the sampled waveform. The predictors of the generated waveform are then compared to the original speech parameters. Any variations within certain predefined tolerances are considered as acceptable, while variations beyond those tolerances are adjusted by revising the synthesis parameters. Thus, a popular term associated with hybrid coding is synthesis by analysis. Two of the more popular implementations of hybrid coding are GSM and a family of codebook excited linear prediction (CELP) methods.

Global System for Mobile (GSM)

One of the most popular examples of hybrid coding is the Global System for Mobile (GSM) communications. GSM is employed for digital cellular operations in Europe and uses a hybrid coding method known as Regular Pulse Excited (RPE) coding which enables voice conversations to be transmitted at 13 kbps. Under RPE coding voice samples are split into frames 20 ms in length. For each frame a set of eight short-term predictor coefficients are obtained. Each frame is then further divided into four 5 ms subframes. The encoder determines a delay and gain for the long-term predictor for each subframe. A total of 40 samples is divided into three possible excitation sequences of 13 samples, and the sequence with the highest energy level is then selected as the best representation of the 20 ms sample.

Code Excited Linear Prediction (CELP)

Although Code Excited Linear Prediction (CELP) dates from the 1980s, it was not until the early 1990s that digital speech processors (DSPs) with sufficient processing capability were developed to implement theory. The development of 200–300 MHz DSPs was accomplished by the development of several versions of CELP.

As its name implies, CELP is based upon the construction of a codebook of excitations. The codebook is adaptively constructed and synthesis parameters resulting from voice samples are matched against entries in the adaptively constructed table. This enables the index of the code to be transmitted instead of actual speech parameters, significantly reducing the quantity of transmission. The original version of CELP, which operates at 16 kbps, was standardized by the ITU in 1992. Although CELP was a considerable improvement in bandwidth utilization over PCM and ADPCM, the size of the speech period and the method used to create a codebook resulted in a relatively long delay, which made this technique unsuitable for use on packet networks whose routing delays introduce additional latency. Recognizing this problem resulted in the development of a low-delay version of CELP referred to as Low-Delay Code Excited Linear Prediction (LD-CELP).

Low-Delay Code Excited Linear Prediction (LD–CELP)

Low-Delay Code Excited Linear Prediction (LD-CELP) provides a high quality of speech reproduction using a 16 kbps encoding rate and a relatively low encoding delay. To accomplish this, LD-CELP uses a 10 ms frame length which minimizes the delay time. LD-CELP was standardized by the ITU as recommendation G.728.

Conjugate-Structure Algebraic Code Excited Linear Prediction (CS-ACELP)

Conjugate-Structure Algebraic Code Excited Linear Prediction (CS-ACELP) provides a high quality of speech reproduction through the use of a series of coding methods. Those methods include interframe correlation preselection of the codebook structure, and the use of a conjugate structure. CS-ACELP was standardized by the ITU as recommendation G.729 and provides an 8 kbps voice digitization operating rate.

In addition to LD-CELP and CS-ACELP, there are several other versions of CELP that warrant discussion. One recently promulgated ITU standard, Recommendation G.723.1, defines a low bit-rate voice compression method which includes a G3 fax transmission capability. Referred to as Multi-Pulse Maximum Likelihood Quantization (MP-MLQ), this speech compression method is combined with ACELP to provide selectable voice digitization rates of 5.3 and 6.3 kbps. By the late 1990s G.723.1 was being implemented by a number of vendors for inclusion in FRADs and routers to provide a voice over Frame Relay and voice over IP networking capability. When used to transmit digitized voice over packet networks, the voice digitization rate of 5.3 and 6.3 kbps results in packet headers being added to transport digitized speech. Depending upon the method used by FRADs and routers to fragment portions of speech into small segments for transport over the packet network, protocol overhead can result in the actual bandwidth consumed increasing by 10–25%. When ACELP/MP-MLQ compression is employed by multiplexers, each voice channel is commonly transported in an 8 kbps multiplexer time slot between geographically separated locations connected by a private network. Another version of CELP, referred to as Enhanced CELP (E-CELP), was being marketed by one vendor as a proprietary version of CELP during 1998. This version of CELP operates at 2400 and 4800 bps.

Networking with the CELP family

Today an expanding base of equipment is reaching the market that typically supports two or more versions from what is essentially a family of CELP algorithms. Equipment such as FRADs, routers, voice servers and multiplexers is obtaining a CELP voice compression capability. In doing so, the use of CELP significantly reduces bandwidth required for voice calls by 50–87%. In addition to saving bandwidth, the use of CELP enables the support of more voice calls on existing transmission facilities or the transportation of calls over networks originally designed to transport data, such as the Internet and public Frame Relay networks. Both of these actions can result in the transportation of a voice call at a fraction of the cost of using the PSTN or even the use of private voice networks that were constructed using PCM and ADPCM. Later in this book, when we turn our attention to digital networking, we will examine how we can construct very economical private voice transporting networks that also transport data.

In concluding this chapter we will turn our attention to a form of voice multiplexing which many equipment vendors implement. Referred to as digital speech interpolation (DSI), this technique can be associated with any of the previously discussed voice digitization methods covered in this chapter. As you might expect, DSI was first employed with PCM and ADPCM, since waveform coding was the earliest method of voice digitization used by communications. More recently, vendors of FRADs, routers and multiplexers that incorporate support for one or more versions of CELP began to add support for DSI.

3.4 DIGITAL SPEECH INTERPOLATION (DSI)

Digital speech interpolation (DSI) can be considered as a form of statistical voice multiplexing, since it takes advantage of the idle moments spread throughout a normal telephone conversation. In this voice digitization method, only active channels are digitized and transmitted, permitting the equipment to take advantage of the half-duplex nature of voice conversations, as well as the natural pauses inherent in speech.

DSI efficiency is based upon the fact that, for a large group of active channels, the long-term idle time per channel is 60–65% of the call duration. The interpolation process operates by filling the gaps in some channels with speech content simultaneously present in other channels.

The ratio of user channels to voice transmission channels is known as the DSI gain. The DSI gain can be as high as 2.5 to 1 for a very large group of user channels. However, for T- and E-carrier circuits that have a relatively small number (24 or 30) of voice channels, when PCM is used for voice digitization, a DSI gain of 2 to 2.2 is used in multiplexer voice modules. Since a speaker's activity is not predictable, the DSI technique becomes more efficient as the number of conversations increases.

Figure 3.12 illustrates an example of the DSI encoding process. Each active channel, denoted by the greater than (>) sign, is digitized and placed into a transmission block. The first field in the block, which is labeled interpolation control, denotes the active channel numbers or addresses of the digitized voice samples that follow in the block. In this example, the interpolation control field would contain channel addresses 3, 5,..., 96.

Originally, most DSI techniques used either PCM, ADPCM or VQL digitization, resulting in the effective data rate per voice channel being reduced to as low as 16 kbps. This data rate

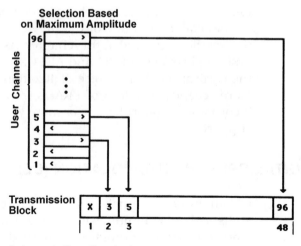

X = Interpolation Control

Figure 3.12 DSI encoding process. The interpolation control header contains the address of the active channels that are encoded and contained in the transmission block

assumes a DSI gain of two. Thus, the use of ADPCM or VQL results in an effective voice digitization rate of

$$\frac{32\,000\,\text{bps}}{2\,\text{DSI gain}} = 16\,000\,\text{bps}$$

Based upon the preceding, the use of DSI with PCM, ADPCM or VQL digitization permits up to 96 voice channels to be carried on a North American T-carrier or 120 channels on a European E-carrier transmission facility.

One of the key advantages of DSI is its ability to provide a 4:1 ratio over the capacity of a T-carrier carrying PCM encoded voice when DSI is used in conjunction with ADPCM or VQL. When a hybrid coding method is used, most multiplexers transport the resulting digitized voice data stream within an 8 kbps or 16 kbps time slot derived from treating the T1 or E1 circuit as a composite data stream instead of a series of 64 kbps time slots. When DSI is employed in conjunction with a waveform coding method, such as a version of CELP, the number of voice channels that can be transported over a T1 or E1 transmission facility substantially increases. For example, consider the use of a version of CELP transported as an 8 kbps data stream instead of PCM's 64 kbps. Without the use of DSI this voice digitization method enables the support of 24×8 or 192 voice channels on a T1 transmission facility and 30×8 or 240 voice channels on an E1 transmission facility. When DSI is used with a 2:1 gain it becomes possible to support 384 channels on a T1 circuit and 480 channels on an E1 circuit, explaining why during the late 1990s several long-distance communications carriers were constructing an infrastructure based upon the use of various hybrid coding techniques to transport voice calls.

To understand the major disadvantages of DSI, let us assume we are transporting voice digitized by ADPCM or VQL, and also assume that in a 6-ms time period there are more than 44 user channels with speech energy that meet the employed selection criteria. This will result in some channels not being transmitted, causing a random 6-ms discontinuity or dropout. This dropout, which results in a cutoff of speech energy, is more commonly known as clipping.

Dropouts are more likely to occur in weak signals since the DSI selection algorithm normally favors high-level signals. Sometimes, dropouts of 18 or 24 ms can be inaudible; however, the human ear can normally detect dropouts in excess of 36 ms.

To prevent clipping or extended dropouts, DSI employs an algorithm to distribute small dropouts evenly over the population of active channels. Unfortunately, this would cause havoc to digital data carried by a DSI system. Thus, to protect data, DSI devices assign a channel carrying modem or digital data traffic to a fixed slot. This shields the data from the interpolation process and ensures that dropouts will not occur; however, it also decreases the overall efficiency of a DSI system.

REVIEW QUESTIONS

1 Discuss the function of each of the three steps in the pulse code modulation process.

2 Why do pulse amplitude modulated pulses require quantization?

3 What is quantization noise and how does it occur?

4 What are two advantages associated with the PCM word encoding form illustrated in Figure 3.4?

5 What is the purpose of a compander?

6 Explain how ADPCM operates at 32 kbps and the advantages associated with this data rate.

7 How many voice conversations can be transported on a T1 carrier using PCM and 32 kbps ADPCM?

8 Discuss several disadvantages associated with the use of ADPCM.

9 What is the major problem associated with CVSD voice digitization that precludes its use by communications carriers?

10 What voice digitization technique provides a measure of protection to the encoded signal that protects it from a small impairment?

11 Discuss an advantage and a disadvantage associated with the use of a vocoding technique, such as linear predictive coding.

12 How does hybrid coding obtain a low bit rate and better reconstructed voice quality than a coding voice digitization technique?

13 Assume you plan to use a multiplexer which supports a version of CELP voice digitization that produces a data stream operating at 16 kbps. What is the maximum number of voice conversations that can be transported when the multiplexer is connected to a T1 circuit?

14 Explain why the efficiency of digital speech interpolation significantly decreases as data traffic transported on a T1 or E1 circuit increases.

4

THE COMMUNICATIONS CARRIER TRANSMISSION INFRASTRUCTURE

The purpose of this chapter is to become acquainted with the evolving communications carrier digital transmission infrastructure that provides the highway used to move voice, data and video between cities and countries. Information presented in this chapter will explain the relationship between different types of copper- and optical-based transmission systems used by communications carriers. Since the different types of digital transmission systems that organizations can obtain to access the communications carrier infrastructure depend upon the manner by which the infrastructure is created, this chapter also forms the foundation for the next chapter in this book. That is, in Chapter 5 we will turn our attention to different types of digital transmission facilities that organizations can employ to access the communications carrier transmission infrastructure.

When examining the communications carrier transmission infrastructure we can classify its evolution into two general periods. The first period represents the use of copper-based access lines to central offices and copper trunks and digital microwave transmission between carrier central offices. Taking writer's liberty, I will use the title 'The Pre-optical Transmission Hierarchy' for the first section in this chapter. Then, the remainder of this chapter will be focused upon obtaining an overview of the communications carrier transmission infrastructure created through the use of transmission on optical fiber. Since the two fiber optic-based transmission systems used by communications carriers are SONET and SDH, we will examine their general transmission characteristics in the second section of this chapter.

4.1 THE PRE-OPTICAL TRANSMISSION HIERARCHY

In this section we will examine the evolution of transmission system facilities used by communications carriers to interconnect their central offices prior to the introduction of SONET and SDH. In doing so we will first focus our attention upon the use of frequency division multiplexing (FDM), which was the earliest technique used to enable multiple simultaneous voice conversations to be routed onto a common circuit interconnecting two carrier central offices. This examination of the use of FDM equipment will be followed by a short review of the operation of time division multiplexing (TDM) equipment whose utilization contributed directly to the development of T- and E-carrier transmission facilities. Based upon the preceding, we will conclude this section by examining the evolution of T-carrier systems and their use by communications carriers to interconnect their central offices.

To reduce the number of physical lines required to connect telephone company offices to one another, communications carriers employ multiplexing. Originally, frequency division multiplexing was used exclusively by communications carriers. Gradually, FDM was replaced by time division multiplexing that utilizes the T- and E-carrier as a digital transport mechanism. This evolution to TDM equipment and T- and E-carrier transmission facilities was based upon the advantages associated with digital signaling in comparison to analog signaling as previously covered in this book.

Frequency division multiplexing

Frequency division multiplexing (FDM) represents the first method developed to enable multiple simultaneous voice conversations to be transported on a trunk used to interconnect communications carrier central offices. FDM was developed prior to the introduction of digital technology and today, for the most part, represents a dated technology. Although almost all FDM equipment has been removed from use by most modern communications carriers, such equipment is still in use in some less developed areas on our globe.

Employing frequency division multiplexing between carrier central offices requires the use of a communications circuit that has a relatively wide bandwidth. This bandwidth is then divided into subchannels by frequency. When a communications carrier uses FDM for the multiplexing of voice conversations onto a common circuit, the 3 kHz passband of each conversation is shifted upward in frequency by a fixed amount of frequency. This frequency shifting places the voice conversation into a predefined channel of the FDM multiplexed circuit. At the opposite end of the circuit, another FDM demultiplexes the voice conversations by shifting the frequency spectrum of each conversation downward in frequency by the same amount of frequency by which it was previously shifted upward.

As previously mentioned, the primary use of FDM equipment by communications carriers was to enable those carriers to carry a large number of simultaneous voice conversations on a common circuit routed between two carrier offices. The actual process for allocating the bands of frequencies to each voice conversation was standardized by the ITU. ITU FDM recommendations govern the channel assignments of voice multiplexed conversations based upon the use of 12, 60 and 300 derived voice channels.

ITU FDM recommendations

The standard group as defined by ITU recommendation G.232 occupies the frequency band from 60 to 108 kHz. This group can be considered as the first level of frequency division multiplexing and contains 12 voice channels, with each channel occupying the 300 to 3400 Hz spectrum shifted in frequency.

The standard supergroup as defined by ITU recommendation G.241 contains five standard groups, equivalent to 60 voice channels. The standard supergroup can be considered as the second level of frequency division multiplexing and occupies the frequency band from 312 to 552 kHz.

The third ITU FDM recommendation, known as the standard mastergroup, can be considered as the top of the FDM hierarchy. The standard mastergroup contains five supergroups. Since each supergroup consists of 60 voice channels, the mastergroup contains a total of 300 voice channels. The standard mastergroup occupies the frequency band from 812 to 2044 kHz. Figure 4.1 illustrates the three standard ITU FDM groups, as well as the relationship between groups.

Time division multiplexing

The development of digital technology made it possible to design equipment that would enable the transmission capacity of a circuit to be shared by time. The technique developed is referred to as time division multiplexing (TDM).

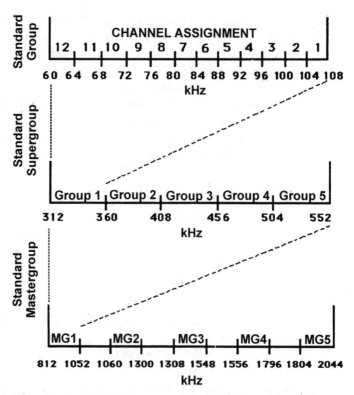

Figure 4.1 Standard ITU FDM groups. ITU FDM recommendations govern the assignment of 12, 60 and 300 voice channels on wideband analog circuits

The fundamental operating characteristics of a TDM are shown in Figure 4.2. Here, each low to medium speed digital data source is connected to the multiplexer through an input/output (I/O) channel adapter. The I/O adapter provides the buffering and control functions necessary to interface low to medium speed data sources to the multiplexer. Within each adapter, a buffer or memory area exists which is used to compensate for the speed differential between the data sources and the multiplexer's internal operating speed. Data is shifted from the terminal devices transmitting to the I/O adapter at different rates, depending upon the speed of the connected input data sources; but when data is shifted from the I/O adapter to the central logic of the multiplexer, or from central logic to the composite adapter, it is at the much higher fixed rate of the TDM. On output from the multiplexer to each data source the reverse is true, since data is first transferred at a fixed rate from central logic to each adapter and then from the adapter to the attached device at the data rate acceptable to the device.

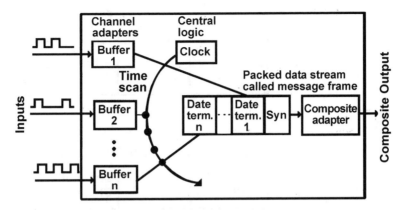

Figure 4.2 Time division multiplexing. In time division multiplexing, data is first entered into each channel adapter buffer area at a transfer rate equal to the device to which the adapter is connected. Next, data from the various buffers are transferred to the multiplexer's central logic at the higher rate of the device for packing into a message frame for transmission

The central logic of the TDM contains controlling, monitoring and timing circuitry which facilitates the passage of individual terminal data to and from the high speed transmission medium. The central logic will generate a synchronizing pattern which is used by a scanner circuit to interrogate each of the channel adapter buffer areas in a predetermined sequence, blocking the bits of characters from each buffer into a continuous, synchronous data stream which is then passed to a composite adapter. The composite adapter contains a buffer and functions similar to the I/O channel adapter. However, it now compensates for the difference in speed between the high speed transmission medium and the internal speed of the multiplexer.

TDM techniques

Two popular TDM techniques are bit interleaving and character interleaving. Bit interleaving is generally used in systems which service synchronous devices, whereas character interleaving is generally used to service asynchronous devices. When interleaving is accomplished on a bit-by-bit basis, the multiplexer takes one bit from each channel adapter and then combines them as a word or frame for transmission. As shown in Figure 4.3A, this technique produces a frame containing one data element from each channel adapter.

When interleaving is accomplished on a character-by-character basis, the multiplexer assembles a full character from each data source into a frame for transmission as shown in Figure 4.3B.

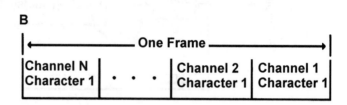

Figure 4.3 TDM techniques. In bit interleaving, a frame is assembled by the TDM gathering one bit from each input data source. In character interleaving, the multiplexer assembles the frame by gathering one character from each input data source: A bit interleaving; B character interleaving

In general, bit interleaving is more efficient than character interleaving as it minimizes the potential delay in the data flow of multiplexed synchronous events. Unfortunately, bit interleaving cannot be used with some types of digital networking activities, such as digital access cross connect (DACC) systems, which are described later in this book and which operate on an eight-bit-per-character basis. To compensate for this dilemma, some multiplexer vendors introduced equipment that operates on a combined bit and character interleaving basis. That is, a portion of the composite bandwidth, or more accurately a portion of the multiplexing frame, is reserved for character interleaving while the remainder of the multiplexer frame is reserved for bit interleaving.

Now that we have an appreciation for the general operation of time division multiplexing, let's turn our attention to the evolution of the T- and E-carriers whose formation is based upon TDM.

T- and E-carrier evolution

T- and E-carrier transmission facilities were originally developed by telephone companies as a mechanism to relieve heavy loading on interexchange circuits. First employed in the 1960s for intra-carrier communications, T- and E-carrier facilities only became available to the general public within the last 15 years as a commercial offering.

Since the mid-1980s, reductions in T- and E-carrier tariffs have made them into an economic data plus voice transportation highway. Currently, T- and E-carrier multiplexers are exhibiting a sales growth of 15–20% per year, among the highest rate of growth of all categories of communications equipment.

The first T-carrier was placed into service by American Telephone & Telegraph in 1962 to ease cable congestion problems in urban areas. Known as T1 in North America, this wideband digital carrier facility operates at a 1.544 Mbps signaling rate since 8000 framing bits per second are included in the signal for synchronization and other functions. Later in this chapter we will briefly examine the framing structure of T1 and E1 transmission systems. In Chapter 6 we will examine in detail different framing methods used on T- and E-carrier transmission facilities.

The term T1 was originally defined by AT&T and referred to 24 64-kbps PCM voice channels carried in a 1.544 Mbps wideband signal. When AT&T initiated use of its T1 carrier, the company employed digital channel banks which were used to interface the analog telephone network to the T1 digital transmission facility.

Channel banks

Channel banks used by telephone companies were originally analog devices. They were designed to provide the first step required in the handling of telephone calls that originated in one central office, but whose termination point was a different central office. The analog channel bank included frequency division multiplexing equipment, permitting it to multiplex, by frequency, a group of voice channels routed to a common intermediate or final destination over a common circuit. This method of multiplexing was previously illustrated in Figure 4.1.

The development of pulse code modulation resulted in analog channel banks becoming unsuitable for use with digitized voice. AT&T then developed the D-type channel bank which actually performs several functions in addition to the time division multiplexing of digital data.

The first digital channel bank, known as D1, contained three key elements as illustrated in Figure 4.4. The codec, an abbreviation for coder–decoder, converted analog voice into a 64 kbps PCM encoded digital data stream. The TDM multiplexes 24 PCM encoded voice channels and inserts framing information to permit the TDM in a distant channel bank to be able to synchronize itself to the resulting multiplexed data stream that is transmitted on the T1 span line. The

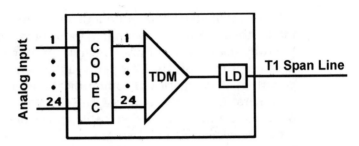

Figure 4.4 The D1 channel bank: TDM = time division multiplexer, LD = line driver

line driver conditions the transmitted bit stream to the electrical characteristics of the T1 span line, ensuring that the pulse width, pulse height and pulse voltages are correct. In addition, the line driver converts the unipolar digital signal transmitted by the multiplexer into a bipolar signal suitable for transmission on the T1 span line. Due to the operation of the digital channel bank, this equipment can be viewed as a bridge from the analog world to the digital world. To ensure the quality of the resulting multiplexed digital signal, AT&T installed repeaters at intervals of 6000 feet (1830 m) on copper span lines constructed between central offices. Although repeaters are still required on local loops to a subscriber's premises and on copper wire span lines, the introduction of digital radio and fiber optic transmission has added significant flexibility to the construction and routing of T-carrier facilities.

Improvements made to the encoding method used by the D1 channel bank resulted in the development and installation of the D1D and D2 channel banks. Channel banks currently used include D3, D4 and D5, whose simultaneous voice channel carrying capacity varies from 24 to 96 channels. Additional information concerning the operation of channel banks is contained in Chapter 6.

Today, T1 lines are available from a variety of communications carriers, including AT&T, MCI, Sprint and others. In Europe, the equivalent T1 carrier, which is known as E1 and CEPT PCM-30, is available in most countries under different names. As an example, in the United Kingdom E1 service is marketed under the name MegaStream.

Framing structure overview

In North America, the T1 carrier was designed to support the transmission of 24 channels of digitized voice.

Each channel is sampled 8000 times per second and eight bits are used to represent the encoded height of the sampled analog wave. For synchronization, as well as other functions that are discussed in Chapter 5, one framing bit is added to the digitized multiplexed data that represents 24 PCM encoded voice conversations.

Thus,

$$8 \text{ bits} \times 24 \text{ channels} + 1 \text{ framing bit} = 193 \text{ bits/frame}$$

Since 8000 frames are transmitted each second, the bit rate is

$$193 \text{ bits/frame} \times 8000 \text{ frames/second} = 1.544 \text{ Mbps}$$

which is the operating rate of the North American T1 carrier facility.

In Europe, the T1-carrier is commonly referred to as an E1 facility or a CEPT PCM-30 system, where CEPT is an acronym for the Conference of European Posts & Telecommunications, a European standards organization.

CEPT uses a 32-channel system, where 30 channels are used to transmit digitized speech received from incoming telephone lines, while the remaining two channels are used for signal and synchronization information. Each channel is assigned a time slot as listed in Table 4.1.

The frame composition of an E1 or CEPT system consists of 32 channels of eight bits per channel, or 256 bits per frame. No framing information is required to be added to the frame as in the North American T1-carrier since synchronization is carried separately in time slot zero.

Since 8000 frames per second are transmitted, the bit rate of an E1 facility becomes

$$256 \text{ bits/frame} \times 8000 \text{ frames/second} = 2.048 \text{ Mbps}$$

Similar to the hierarchy illustrated for FDM in Figure 4.1, communications carriers developed a hierarchy of digital carrier levels. Table 4.2 lists the digital carrier hierarchy levels in North America, Europe and Japan.

Table 4.1 CEPT time slot assignments.

Time slot	Type of information
0	Synchronization (framing)
1–15	Speech
16	Signaling
17–31	Speech

Table 4.2 Digital hierarchy levels.

North America

Line type	Line signal standard	Number of voice circuits	Bit rate (Mbps)
T1	DS1	24	1.544
T1C	DS1C	48	3.152
T2	DS2	96	6.312
T3	DS3	672	44.736
T4	DS4	4032	274.176
T5	DS5		
T6	DS6		

Europe

Level number	System	Number of voice circuits	Bit rate (Mbps)
1	M1	30	2.048
2	M2	120	8.448
3	M3	480	34.368
4	M4	1920	139.264
5	M5	7680	565.148

Japan

Level number	System	Number of voice circuits	Bit rate (Mbps)
1	F-1	24	1.544
2	F-6M	96	6.312
3	F-32M	480	34.064
4	F-100M	1440	97.728
5	F-400M	5760	397.20
6	F-4.6G	23040	1588.80

In North America, T1C and T2 were developed to boost the carrying capacity of copper wire pairs beyond that obtained by using T1. These facilities are primarily restricted to use by telephone companies. T3, which operates at 44.736 Mbps, initially was offered to commercial organizations during 1988. By 1998 T3 was among the most popular types of digital transmission facilities used by Internet Service Providers (ISPs) to connect their network to a peering point operated by a Network Service Provider (NSP) that operates the backbone network on the Internet. In addition, T3 transmission facilities were also being used by individual organizations with popular World Wide Web server locations as a connection mechanism to an ISP or directly to an NSP as well as by organizations for the construction of high speed private networks designed to transport voice, data and video.

The line types listed in the top portion of Table 4.2 actually reference the type of signal each line is capable of carrying. A T1 line carries a DS1 signal. Here DS1 (digital signal, level 1) is the 1.544 Mbps signal defined by AT&T to include pulse height and width, impedance and other parameters. Although column two, headed 'Line signal standard', commences with DS1, in actuality the lowest signal level in the digital hierarchy is DS0 (digital signal, level 0). DS0 refers to each 64 kbps digitized PCM data stream generated in a D-type channel bank, with 24 such DS0 channels along with framing bits used to form a DS1 signal.

Figure 4.5 illustrates the North American digital signal hierarchy. Note that DS0 originates at the digital channel bank located in the lower left portion of the illustration. The five data rates shown entering the data multiplexer in the upper left portion of Figure 4.5 are Dataphone Digital Services transmission facilities, which are also commonly referred to as subrate services as they operate below the DS1 rate. Dataphone Digital Services was originally developed by AT&T as an all-digital transmission facility for the transmission of data as opposed to voice. Since its introduction by AT&T, other communications carriers have introduced equivalent digital facilities for the transmission of data. The characteristics and operation of AT&T's DDS transmission facility, as well as other digital services, are discussed in Chapter 5.

In the North American digital signal hierarchy, the M12 multiplexing system used by AT&T accepts four DS1 input signals and produces a DS2 output representing 96 DS0 channels operating at 6.312 Mbps. The AT&T M13 multiplexer operates upon 28 DS1 inputs, while that carrier's M23 multiplexer operates upon seven DS2 inputs, with both devices generating a DS3 output operating at 44.736 Mbps which represents 672 DS0 channels. The highest order AT&T multiplexer, which is the M34 device, accepts six DS3 inputs to form one DS4 output operating at 274.176 Mbps which represents 4032 DS0 channels.

4.2 SONET AND SDH

During the late 1970s and early 1980s communications carriers recognized the advantages of the high bandwidth and immunity to electrical disturbances associated with the use of optical transmission by installing a series of lightwave transmission systems.

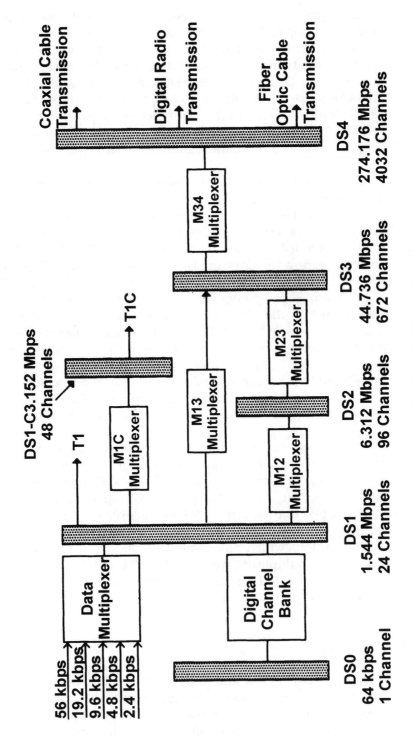

Figure 4.5 North American digital signal hierarchy

Although such optical systems were successful with respect to their use, enabling high speed transmission at very low data rates, each system was proprietary. This resulted in compatibility and interoperability problems occurring between different communications carriers. At first such problems represented more of a minor inconvenience since the requirement for a very high speed transmission facility was in its infancy, and in the United States divestiture of the Bell system was in its planning stage. However, divestiture became a reality, which resulted in a requirement to standardize optical transmission methods and equipment operations to promote interoperability between what became many independent communications carriers. An added virtue of standardization was the fact that it enabled communications carriers to mix and match equipment from different suppliers, permitting competition and the financial benefits from having an open optical transmission system.

Standards evolution

Work on a standard for optical transmission was begun in 1984 by the Exchange Carriers Standards Association (ECSA). The resulting standard, referred to as the Synchronous Optical NETwork (SONET), defines a hierarchy of synchronous transport signals as well as a synchronous multiplexing format for carrying lower level digital signals, such as DS1 and DS3 over a fiber-optic based transmission hierarchy. The development of the SONET standard was followed by the effort of the ITU to define an optical transmission standard to address the differences between North American and European transmission hierarchies. That effort resulted in the publication of the Synchronous Digital Hierarchy (SDH) standard by the ITU during 1989. SDH represents a world standard, while SONET, which predates SDH, can be considered as a subset of SDH, with the two standards actually converging at a 155.520 Mbps data rate.

Clocking

SONET and SDH transmission hierarchies include an architecture in which pointers are used to position signals within their frames. This pointing technique, as well as the use of a negative stuffing byte to compensate for clocking differences, is described in Chapter 6. In SONET and SDH the average frequencies of all clocks are either the same (synchronous) or nearly the same, and are described by the term plesiochronous. The key advantage associated with this type of clocking is the fact that many separate optical signals can be grouped together via a standard for transmitting at much higher data rates than those commonly available in asynchronous networks. Here the term 'asynchronous' refers to the fact that there is a relatively large difference in the clocks used to define signaling between two devices. In fact, signals such as DS1s and DS3s are considered to be asynchronous due to clock differences between two devices on a T1 or T3 circuit. This resulted in the requirement to perform what is referred to as 'bit stuffing' when DS1, DS3 and other asynchronous signals are multiplexed. In comparison, SONET and SDH were designed to perform multiplexing without requiring bit stuffing, which also facilitates demultiplexing. This clocking advantage of SONET and SDH will become clearer as we proceed through subsequent chapters in this book.

The optical hierarchy

SONET transmission rates begin at 51.84 Mbps and increase to 4.97 Gbps in basic increments of 51.84 Mbps. In comparison, the SDH basic operating rate is 155.52 Mbps which is equivalent to SONET's STS-3 (Synchronous Transport Signal Level 3) level. Table 4.3 illustrates the relationship between SONET and SDH transmission hierarchies.

In examining the entries in Table 4.3 it is important to note that SDH commences at an operating rate of 155.52 Mbps and that there is no such designation as 'Synchronous Transport Modulo 0' or STM-0. STM-0 is an author liberty taken to indicate that the base level for SDH is equivalent to three of SONET's Synchronous Transport Signal Level 1 (STS-1) signals, each of which would be equivalent to an STM-0 if SDH's base level was 51.84 Mbps instead of 155.52 Mbps. Note that for both SONET and SDH, higher level signals have bit rates which correspond to multiples of their STS-1 and STM-0/STM-1 signals. This facilitates the multiplexing of multiple lower order signals onto a higher level signal via the use of byte multiplexing, without requiring bit stuffing which both adds to the complexity of the multiplexing process and makes it difficult to directly remove DS0s from a multiplexed data stream. The complexity associated with multiplexing individual DS1 signals and the reason why the bit stuffing process adds to the complexity of the multiplexing process will become clearer as we examine the

Table 4.3 The SONET/SDH Transmission Hierarchies.

Fiber-Optic signal (OC-level)	Electrical signal (STS/STM)*	Bit rate (Mbps)	SONET capacity	SONET buildup	SDH buildup	SDH capacity
OC-1	STS-1/STM-0	51.840	28 DS1s or 1 DS3	—	—	21 E1s
OC-3	STS-3/STM-1	155.520	84 DS1s or 3 DS3s	3XSTS-1	3XSTM-0	63 E1s or 1 E4
OC-12	STS-12/STM-4	622.080	336 DS1s or 12 DS3s	12XSTS-1	4XSTM-1	252 E1s or 4 E4s
OC-48	STS-48/STM-16	2488.320	1344 DS1s or 48 DS3s	48XSTS-1	16XSTM-1	1008 E1s or 16 E4s
OC-192	STS-192/STM-64	9953.280	5376 DS1s or 192 DS3s	192XSTS-1	64XSTM-1	4032 E1s or 64 E4s

*STS: Synchronous Transport Signal; STM: Synchronous Transport Module.

operation of copper-based digital networking facilities in Chapter 5 and T- and E-carrier framing and coding formats in Chapter 6.

REVIEW QUESTIONS

1 Under frequency division multiplexing how do two or more conversations share the use of a common transmission facility?

2 Under time division multiplexing how do two or more conversations share the use of a common transmission facility?

3 What is the difference between a bit interleaving TDM and a character interleaving TDM? When would you prefer to use a character interleaving TDM?

4 What are the basic functions of the three key components in a digital channel bank?

5 What is the relationship between the operating rate of a T1 carrier and its framing rate?

6 Where is framing information conveyed on an E1 transmission facility?

7 Describe the driving forces behind the development of SONET and SDH.

8 What is the difference between asynchronous and plesiochronous signaling with respect to clocking?

9 Why is the multiplexing of SONET and SDH signals easier than the multiplexing of T- and E-carrier signals?

5

COPPER-BASED DIGITAL NETWORKING FACILITIES

In this chapter we will examine the operation and utilization of several types of North American and European digital network facilities that were originally developed when the communications carrier network infrastructure was based upon the use of copper and digital microwave transmission. Although the current communications carrier network infrastructure is primarily based upon the use of fiber optics, the previously developed transmission facilities were integrated into the optical hierarchy and remain readily available for organizations to use.

In this chapter we will first focus our attention upon obtaining an appreciation for the relationship between different types of digital transmission services and public and private network services, such as X.25, Frame Relay, Switched Multimegabit Digital Service (SMDS) and ATM. Once this is accomplished we will turn our attention to subrate facilities which can be defined as a digital network facility operating at or below the DS0 data rate of 64 kbps. Since a grouping of 24 or 32 DS0 channels was originally multiplexed onto a North American or European E-carrier facility, our next focus of attention will be upon T- and E-carrier transmission facilities. This will be followed by an overview of the method used by communication carriers to switch DS0 channels between T- and E-carriers, a process known as digital access cross connect (DACS). This overview of DACS will include how it can be used by end-user organizations to control the routing of voice and data within an end-user's network. In concluding this chapter we will examine the operation of the T3 transport facility and fractional T1 and T3 transport facilities which are commonly referred to as FT1 and FT3.

5.1 TRANSMISSION AND NETWORK SERVICES

Although the most common use of copper-based digital transmission facilities involves the use of leased lines to construct private networks, a second significant use is to obtain access to different types of public and private network services. Such services can range in scope from the original X.25 packet network and the more modern Frame Relay packet network and Switched Multimegabit Data Services (SMDS) packet services to a communications carrier's Asynchronous Transfer Mode (ATM) transmission service.

Network service access

Figure 5.1 illustrates the general relationship between different types of transmission services and the use of those services to access

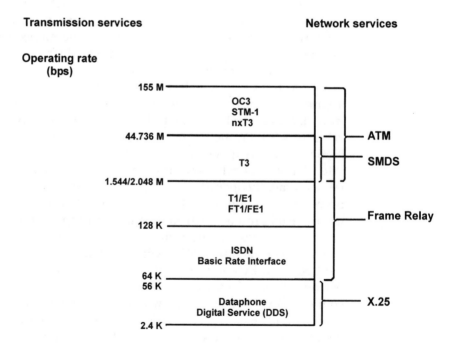

Figure 5.1 Relationship between communications carrier transmission and network services

different types of network services. In examining the relationship between transmission services and network services shown in Figure 5.1 several items warrant an explanation. First, the availability of different transmission services to access different network services can depend considerably on the geographical area. Most communications carriers do not offer transmission facilities at T3 and higher rates in rural areas, nor in certain portions of urban areas, due to the cost associated with installing the transmission facility to a subscriber location. Secondly, even if offered in a certain geographical area, the time required for installation may be sufficiently longer than that for lower operating rate transmission facilities. Thus, you should examine both transmission availability and installation time to acquire an available service.

One transmission service purposely omitted from Figure 5.1 is FDDI. The reason for this omission is the fact that its use by communications carriers is as a general transmission service used to provide access to a carrier's central office for all subscriber transmission requirements. Thus, FDDI could be used to provide access to any or all of the network services shown in Figure 5.1 as well as for local access to different types of leased lines.

5.2 SUBRATE FACILITIES

A subrate facility can be classified as a digital network communications line that operates at a data rate less than a T- or E-carrier data rate — 1.544 Mbps in North America and 2.048 Mbps in Europe. Two commercial examples of digital subrate facilities are AT&T's Dataphone Digital Service and British Telecom's Kilo-Stream service offering.

Dataphone Digital Service (DDS)

Dataphone Digital Service (DDS) was approved by the US Federal Communications Commission in December 1974. Currently, there are over 100 cities in the United States that are connected to the DDS network as well as connections which provide an interconnection capability to digital networks operated by other US and foreign communications carriers.

DDS is an all-synchronous facility. Currently supported transmission rates are listed in Table 5.1. For transmission at different

Table 5.1 DDS offerings.

Data rate (kbps)	Type of service
2.4	leased
4.8	leased
9.6	leased
19.2	leased
56.0	leased
56.0	switched

data rates, specialized equipment to include multiplexers and/or converters must be employed.

Carrier structure

DDS facilities are routed from a subscriber's location to an office channel unit (OCU) located in the carrier's serving central office. Since there is a variety of multiplexing methods that can be employed by a communications carrier to combine DDS facilities onto a T-carrier, let us focus our attention upon two methods that will illustrate the relationship between DDS and a T1 circuit. Figure 5.2 illustrates the multiplexing arrangement within an AT&T serving central office that supports DDS transmission at 9.6 kbps and 56 kbps.

The data service units (DSUs) at the subscriber's location can be viewed as 'digital modems' since they modulate the unipolar signal received from data terminal equipment, including computer ports, multiplexer ports and terminal ports, into a modified bipolar signal suitable for transmission on the DDS network. Originally, separate channel service units (CSUs) and data service units were required to interface equipment to the DDS network. Today, just about all vendors manufacture DSUs that, in effect, combined the functions performed by separate DSUs and CSUs. The operation of both CSUs and DSUs is covered later in this chapter.

The user TDMs shown in Figure 5.2 illustrate two methods by which end-users can transmit data to maximize the data handling capacity of different DDS facilities. The user TDM shown in the upper left corner of Figure 5.2 illustrates how asynchronous transmission can be supported on DDS which is an all-synchronous transmission facility. In this example, 1200 bps asynchronous data sources are multiplexed into a 9.6 kbps synchronous data source for transmission onto DDS via the use of a DSU operating at

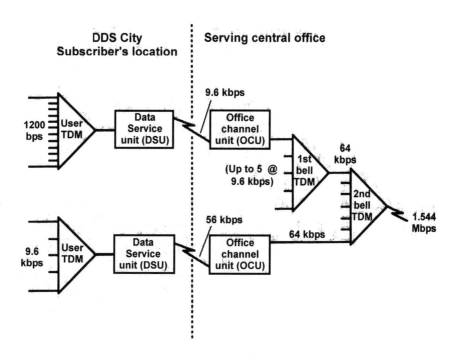

Figure 5.2 DDS multiplexing arrangement

9.6 kbps. In the lower left portion of Figure 5.2, five 9.6 kbps asynchronous or synchronous data sources are multiplexed to obtain a 56 kbps synchronous data rate suitable for transmission on DDS. In both examples, one physical DDS circuit is used to transmit multiple logical channels of data.

The signals from the DSUs are terminated into a complementary office channel unit in the serving central office. From there, they enter into a multiplexing hierarchy which may carry voice as well as data.

Framing formats

One of the more interesting aspects of DDS is the constraints upon its transmission rate resulting from the formats used to encode user data. User data transmitted at 56 kbps is increased to a DS0 64 kbps data rate at the OCU, and that device inserts groups consisting of seven bits of customer data into an eight-bit byte as

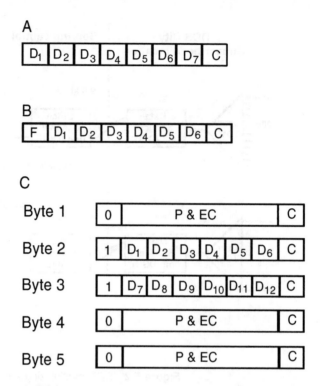

Figure 5.3 DDS framing formats: A 56 kbps: D=data bits, C=control bit; B 2.4, 4.8, 9.6 kbps: F=frame bit, D=data bits, C=control bit; C 19.2 kbps: D=data bits, C=control bit, P & EC=parity and error connection bits

illustrated in Figure 5.3A. In this encoding format the control bit (C) added to every seven bits of customer data is set to a '1' if the byte contains customer data, while a value of '0' indicates that the byte contains network control data, such as idle or maintenance codes or control information. Since a DS0 signal results in the transmission of an eight-bit byte 8000 times per second, this framing format results in 8 kbps of control bits being added to the 56 kbps customer data rate.

The construction of DS0 signals from the 2.4, 4.8 and 9.6 kbps DDS subrates is illustrated in Figure 5.3B. As indicated, customer data is inserted into eight-bit bytes with six bits of user data framed by a frame bit (F) and a control bit (C). Once 2.4, 4.8 or 9.6 kbps DDS data streams are framed, one of two methods is used to place the framed data onto a DS0 channel. When 'byte stuffing' is used, the frame bit is set to '1' and the customer data is repeated the required number of times to create a 64 kbps DS0 signal. Thus, the eight-bit byte containing six bits of user data is repeated at five, 10 and 20 times to enable 9.6, 4.8 and 2.4 kbps DDS data to be placed on a 64 kbps channel. When the F bit is set to '1' the frame format illustrated in Figure 5.3B is referred to as a DS0-A format. Thus, DS0-A data can bypass the first level TDM illustrated in Figure 5.2 and be fed directly into the second level TDM.

The second method of placing 2.4, 4.8 or 9.6 kbps DDS data onto a DS0 channel involves the use of the first level TDM illustrated in Figure 5.2. When this occurs, five 9.6, 10 4.8, or 20 2.4 kbps formatted signals are multiplexed onto a single DS0 channel. To distinguish between the repeating of the same data resulting from byte stuffing and the multiplexing of different DDS signals, the framing bit is altered from all 1s in byte stuffing to a subrate framing pattern to indicate multiplexing of different DDS data sources. When this framing pattern occurs, the resulting framing format is referred to as a DS0-B format. Obviously, DS0-B formatting is more efficient than DS0-A formatting as the latter would require 20 DS0 channels to transmit 20 2.4-kbps signals, while the former would require only one DS0 channel.

The introduction of 19.2 kbps DDS service required a substantial framing format change to accommodate this data rate. As illustrated in Figure 5.3C, five bytes are required to carry 19.2 kbps customer data since three bytes are used for parity and error correction functions. In this framing format the frame bit in each byte (bit 1) results in a '001100' repeating pattern. Twelve bits of customer data are placed into two six-bit groups contained in bytes 2 and 3, resulting in 12 data bits being carried in every five-byte group of 40 bits. Thus, the use of the 64 kbps DS0 channel produces an effective data rate of $64 \times 12/40$, or 19.2 kbps.

Signaling structure

A modified bipolar signaling structure is used on DDS facilities. The modification to bipolar return to zero signaling results in the insertion of zero suppression codes to maintain synchronization whenever a string of six or more zeros is encountered. Otherwise, repeaters on the span line routed between the carrier office and the customer may not be able to obtain clocking from the signal and could then lose synchronization with the signal.

To ensure a minimum 1s density, at 2.4, 4.8, 9.6 and 19.2 kbps any sequence of six consecutive 0s is encoded as 000X0V, where

0 denotes zero voltage transmitted (binary 0),
X denotes a zero or + or − A volts, with the polarity determined by conventional bipolar coding,

V denotes + or − A volts, with the polarity in violation of the bipolar rule.

Figure 5.4 illustrates the zero suppression sequence used to suppress a string of six consecutive 0s. For transmission at 56 kbps, any sequence of seven consecutive 0s is encoded as 0000X0V.

Timing

Precise synchronization is the key to the success of an all-digital network. Timing ensures that data bits are generated at precise intervals, interleaved in time and read out at the receiving end at the same interval to prevent the loss or garbling of data.

To accomplish the necessary clock synchronization on the AT&T digital network, a master clock is used to supply a hierarchy of timing in the network. Should a link to the master clock fail, the nodal timing supplies can operate independently for up to two weeks without excessive slippage during outages. In Figure 5.5, the hierarchy of timing supplies as linked to AT&Ts master reference clock is illustrated. As shown, the subsystem is a treelike network containing no closed loops.

Service units

When DDS was introduced, both a channel service unit (CSU) and a data service unit (DSU) were required to terminate a DDS line.

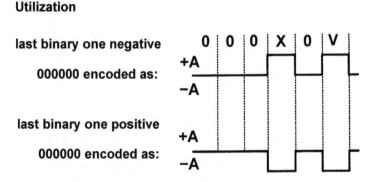

Format: 000X0V

Utilization

last binary one negative

000000 encoded as:

last binary one positive

000000 encoded as:

Figure 5.4 DDS zero suppression sequence

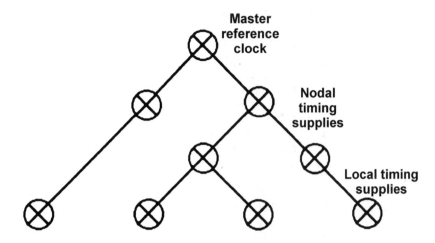

Figure 5.5 DDS timing subsystem

The DSU converts the signal from data terminal equipment into the bipolar format used with DDS.

The DDS master reference clock is an atomic clock that is accurate to 0.01 part per million (ppm). This clock was installed by AT&T at Hillsboro, MO, which is the geographic center of the United States and whose location ensures a minimal variance in propagation delay time between DDS nodes connected to the master reference clock. The master reference clock oscillates at a rate known as the basic system reference frequency and is the most accurate of three timing sources used by digital facilities. The other two sources of timing include channel banks and loop timing where clocking is obtained from a high-speed circuit. A discussion of channel bank and loop timing is contained in Chapter 7.

The DSU interface to the DTE is accomplished by the use of a standard 25-pin EIA RS-232/V.24 female connector on the 2.4 kbps through 19.2 kbps units. The wideband, 56 kbps device utilizes a 34-pin ITU, V.35 (Winchester) female-type connector. Table 5.2. lists the RS-232 and V.35 interchange circuits commonly used by most DSUs. Since DTEs are normally attached to the DSU, the latter's interface is normally configured as data communications equipment (DCE) by the manufacturer.

Prior to deregulation, the CSU was provided by the communications carrier, while the DSU could be obtained from the carrier or from third-party sources. This resulted in an end-user connection to the DDS network similar to that illustrated in the top portion of Figure 5.6 where the CSU terminated the carrier's

Table 5.2 DSU interchange circuits

RS-232 interface (DCE)		V.35 interface (DCE)	
Pin	Signal	Pin	Signal
1	Chassis ground	P	Transmit data (A)
2	Transmit data	S	Transmit data (B)
3	Receive data	R	Receive data (A)
4	Request-to-send	T	Receive data (B)
5	Clear-to-send	C	Request-to-send
6	Data set ready	D	Clear-to-send
7	Signal ground	H	Data terminal ready
8	Carrier detect	E	Data set ready
9	Positive voltage	B	Signal ground
10	Negative voltage	F	Receive line signal detect
15	Transmit clock	Y	Transmit timing (A)
17	Receive clock	AA	Transmit timing (B)
20	Data terminal clock	V	Receive timing (A)
24	External transmit clock	X	Receive timing (B)
		U	External transmit timing (B)
		W	External transmit timing (B)
		L	Local loop-back

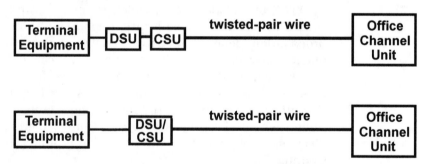

Figure 5.6 DSU/CSU connection

four-wire loop and the DSU was cabled to the CSU. In this configuration the CSU terminates the carrier's circuit. In addition, a separate CSU was designed to perform signal regeneration, monitor incoming signals to detect bipolar violations and perform remote loopback testing. Interfacing between the DSU and CSU is accomplished by the use of a 15-pin female D-type connector which utilizes the first six pins: pin 1 is signal ground, pin 2 is status indicator, pins 3 and 4 are the receive signal pair, while pins 5 and 6 are the transmit signal pair.

Since deregulation, over 50 third-party vendors have introduced combined DSU/CSU devices, integrating the functions of both devices into a common housing which is powered by a common power supply. The lower portion of Figure 5.6 illustrates the connection of end-user terminal equipment to DDS using a combined DSU/CSU unit.

Compression and multiplexing features

Among the more interesting features added to third party integrated DSU/CSU products are a data compression capability that can significantly boost the information transfer rate achievable over a digital transmission facility, and a multiplexing capability which enables multiple data streams to share the transport capability of a common digital transmission facility. One example of a combined DSU/CSU which includes both features is the Motorola 3512 Synchronous Data Compressor (SDC) DSU/CSU. This device can be set to provide an attached terminal device such as a router or multiplexer with a clocking rate up to 256 kbps when used for transmission on a 56 or 64 kbps leased line. Figure 5.7 illustrates an example of the use of a pair of Motorola 3512 SDC DSU/CSUs to obtain a throughput rate up to 256 kbps over a 56 kbps DDS transmission facility. Although the 3512s clock data from the routers at 256 kbps, the actual throughput is variable and depends upon the susceptibility of the data being transmitted to compression. Each 3512 includes a data buffer which fills when the devices' compression engine cannot reduce the input data stream sufficiently to support the ratio of data clocking to the 56 kbps operating rate of the digital circuit. As the buffer fills up to a certain predefined threshold, the 3512 will implement flow control as a mechanism to prevent buffer overflow and the loss of data. To do so the 3512 will temporarily stop its transmit clock and de-assert the Clear-to-Send (CTS) signal. Once the buffer is emptied to a sufficient level, clocking and CTS assertion resumes.

In addition to supporting data compression, the Motorola 3512 includes a built-in limited functioning time division multiplexing capability. The reason its TDM capability is limited results from the design of the DSU/CSU which includes three data ports, of which only one port performs data compression. The two non-compression performing ports were included to support transparent data transfer applications that require a constant throughput and low delay. An example of the latter is the primary line in a transmission group of lines interconnecting two IBM front-end processors. That line is used to perform an initial program load (IPL) into a remote

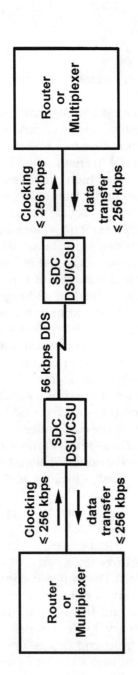

Figure 5.7   Compression-performing integrated DSU/CSUs can boost the throughput obtainable on a 56 kbps DDS leased line

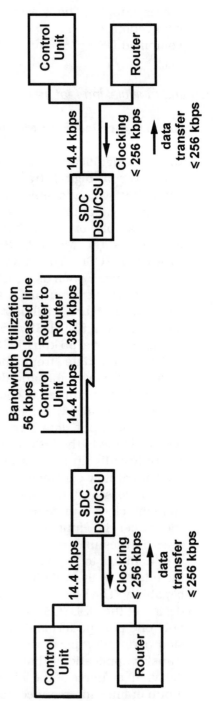

Figure 5.8   Multiplexing and compression data using Motorola integrated DSU/CSUs

front-end processor and typically cannot tolerate delays resulting from a DSU/CSU buffering data so it can apply a compression algorithm.

Figure 5.8 illustrates an example of the use of both multiplexing and data compression. In this example it was assumed that a remote control unit is provided with 14.4 kbps of the bandwidth of the 56 kbps DDS leased line while the routers are provided with 38.4 kbps. Note that, while the 3512 DSU/CSU is shown providing clocking to each router at 256 kbps, the resulting compressed data stream contends for only 38.4 kbps of bandwidth on the 56 kbps DDS circuit. Thus, it may be more practical to lower clocking from the integrated DSU/CSUs to the routers to a lower rate, such as 128 kbps.

DSU/CSU tests and indicators

Through the use of intentional bipolar violations, the DSU/CSU can generate a request to the OCU for the loop-back of the received signal onto the transmit circuit or it can interpret DDS network codes and illuminate relevant indicators on the device. When the loop-back button on the DSU/CSU is pressed, the device will generate four successive repetitions of the sequence 0B0X0V when operating at data rates up to 19.2 kbps or N0B0X0V at 56 kbps, where

B denotes $\pm$A volts, with the polarity determined by bipolar coding for a binary 1,

X denotes a zero volt for coding of binary 0 or B, depending upon the required polarity of a bipolar violation,

V denotes $\pm$A volts, with the polarity determined by the coding of a bipolar violation,

N denotes a don't care condition where the coding for a binary 0 or binary 1 is acceptable.

In Figure 5.9, the DDS loop-back six-bit sequence is illustrated for data rates at or under 19.2 kbps. Note that the sequence transmitted is dependent upon whether the previous binary 1 was transmitted as a positive or negative voltage.

Other bipolar violation sequences used by DDS include an idle sequence which indicates that a DTE does not have data to transmit, an out-of-service sequence and an out-of-frame sequence. The idle sequence is generated by the DSU while the out-of-service and out-of-frame sequences indicate a problem in

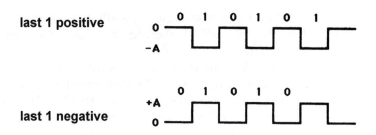

Figure 5.9 DDS loop-back sequence (data rates up to 19.2 kbps). DDS loop-back codes are intentional bipolar violations

the DDS network and are generated by the network and used by the DSU to illuminate an appropriate indicator on the device.

DDS II

During 1988, AT&T introduced a new version of its DDS facility commonly referred to as DDS II. One of the key advantages of DDS II is its capability to provide a diagnostic channel along with the primary subrate channel. This diagnostic channel is obtained through a modification to the framing used by DDS and requires the use of a special DSU/CSU that supports the new channel.

Through the use of DDS II, end-users can perform non-disruptive testing or use the channel for network management purposes. The key to obtaining the ability to derive a secondary channel on DDS is the use of the network control or C bit.

In conventional DDS the C bit, which is bit 8 in each DDS eight-bit byte, is transmitted as a binary 1 whenever a DTE requests access to a channel by turning its request-to-send (RTS) signal on. With the C bit continuously set to a 1 the DTE can transmit an unrestricted stream of data to include continuous zeros since every eight bit will automatically set to a 1. By robbing this bit once every third byte, AT&T established a virtual path for diagnostic use.

The diagnostic channel data rate for 56 kbps DDS II is obtained by multiplying the full DS0 rate of 64 kbps by one-eighth which represents the C bit's portion of the DS0 rate to obtain 8 kbps. Next, since the bit robbing occurs every third byte, the resulting data rate becomes $8000 \times \frac{1}{3}$, or $2666\frac{2}{3}$ bps. Similarly, dividing $2666\frac{2}{3}$ bps by the number of 9.6, 4.8 or 2.4 kbps channels multiplexed onto a DS0 channel results in the diagnostic data rate for DDS II at those data rates. For example, since five 9.6-kbps DDS channels are used to fill a DS0, this means that each 9.6-kbps DDS channel would have a diagnostic channel operating rate of $2666\frac{2}{3}/5$, or $533\frac{1}{3}$ bps. Similarly, since 10 4.8-kbps channels are used to fill a DS0, this means that each 4.8-kbps DDS channel has a diagnostic channel operating at $2666\frac{2}{3}/10$, or $266\frac{2}{3}$ bps. Completing our operating rate examination, since 20 2.4-kbps DDS channels can fill a DS0, each has a diagnostic channel that operates at $2666\frac{2}{3}/20$, or $133\frac{1}{3}$ bps.

Because the diagnostic channel operating rates of DDS II do not represent conventional device operating rates, integrated DSU/CSUs designed to support those secondary channels commonly perform a format and speed conversion to the nearest lower standard ASCII asynchronous data rate. That is, $2666\frac{2}{3}$ bps is typically converted to 2400 bps, $533\frac{1}{3}$ to 300 bps, $266\frac{2}{3}$ bps to 150 bps, and $133\frac{1}{3}$ bps to 75 bps.

Although older DSU/CSU devices can support transmission on DDS II, those devices cannot support the use of the secondary channel capability provided by this modification to DDS. To do so requires the use of newer DSU/CSU devices that support the multiplexing of diagnostic data onto every third C bit position.

Analog extensions to DDS

AT&T provides an 831A auxiliary set which allows analog access to DDS for customers located outside the DDS servicing areas. The 831A connects the EIA RS-232 interfaces between a data service unit and a modem. The 831A contains an eight-bit elastic store, control, timing and test circuits which allow loop-back tests toward the digital network. The elastic store is a data buffer that is required by the DSU to receive data from the modem in time with the modem's receive clock. The data is then held in the elastic store until the DSU's transmit clock requests it. Thus, the buffer serves as a mechanism to overcome the timing differences between the clocks of the two devices. In the reverse direction, no buffer is required when the DSU's receive clock is used as the modem's external transmit clock. When the modem cannot be externally clocked or when one DSU is connected to a second DSU or a DTE that cannot accept an external clock, a second elastic store will be required. Figure 5.10 illustrates a typical analog extension to a DDS servicing area.

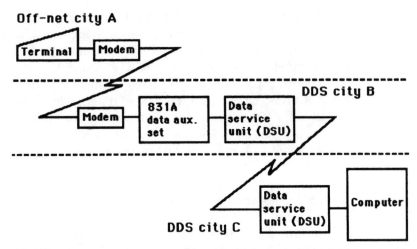

Figure 5.10 Analog extension to DDS. An analog extension to DDS requires the use of one or more elastic stores to compensate for timing differences between a modem and a DSU

British Telecom KiloStream

KiloStream is a point-to-point leased line digital service that was first offered commercially in January 1983. Although KiloStream is similar to DDS, there are several significant differences that warrant discussion.

Like DDS, KiloStream is an all-synchronous facility. British Telecom provides a network terminating unit (NTU) which is similar to DSU/CSU to terminate the subscriber's line. The NTU provides an ITU interface for customer data at 2.4, 4.8, 9.6 or 48 kbps to include performing data control and supervision, which is known as structured data. At 64 kbps, the NTU provides an ITU interface for customer data without performing data control and supervision, which is known as unstructured data.

The NTU controls the interface via ITU Recommendation X.21, which is the standard interface for synchronous operation on public data networks. An optional V.24 interface is available at 2.4, 4.8 and 9.6 kbps, while an optional V.35 interface can be obtained at 48 kbps. The X.21 interface is illustrated in Figure 5.11. Here the control circuit (C) indicates the status of the transmitted information (data or signaling), while the indication circuit (I) signals the status of information received from the line. The control and indication circuits control or check the status bit of an eight-bit envelope used to frame six information bits.

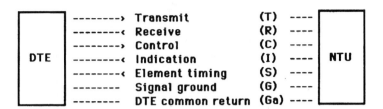

Figure 5.11 CCITT X.21 interface circuits

Data encoding

Customer data is placed into a 6+2 format to provide the signaling and control information required by the network for maintenance assistance. This is known as envelope encoding and is illustrated in Figure 5.12.

The NTU performs signal conversion, changing unipolar non-return to zero signals from the V.21 interface into a di-phase WAL 2 encoding format. This ensures that there is no dc content in the signal transmitted to the line, provides isolation of the electronic circuitry from the line, and provides transitions in the line signal to enable timing to be recovered at the distant end. Table 5.3 lists the NTU operational characteristics of KiloStream.

The KiloStream network

In the KiloStream network, the NTUs on a customer's premises are routed via a digital local line to a multiplexer operating at 2.048 Mbps. This data rate is the European equivalent of the T1 line in the United States that operates at 1.544 Mpbs. The multiplexer can support up to 31 data sources and may be located at the local telephone exchange or on the customer's premises if traffic justifies. It is connected via a digital line or a radio system into the British Telecom KiloStream network as illustrated in Figure 5.13.

Unlike true CEPT, the 2.048 Mbps E1-carrier used for Kilo-Stream used 31 DS0 channels. Normally, CEPT uses one channel

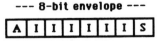

Figure 5.12 KiloStream envelope encoding: A=alignment bit which alternates between '1' and '0' in successive envelopes to indicate the start and stop of each 8-bit envelope, S=status bit which is set or reset by the control circuit and checked by the indicator circuit, I=information bits

Table 5.3 KiloStream NTU operational characteristics.

Customer data rate (kbps)	DTE/NTU interface	Line data rate (kbps)	NTU operation
2.4	X.21	12.8	6+2 envelope encoding
4.8	X.21	12.8	6+2 envelope encoding
9.6	X.21	12.8	6+2 envelope encoding
48	X.21	64	6+2 envelope encoding
64	X.21	64	No envelope encoding
48	X.21 bis/V.35	64	6+2 envelope encoding
2.4	X.21/V.24	12.8	6+2 envelope encoding
4.8	X.21/V.24	12.8	6+2 envelope encoding
9.6	X.21/V.24	12.8	6+2 envelope encoding

for signaling. Since there is no direct voice signaling on KiloStream, time slot 16, which normally would carry that information, can be used for data. From an examination of Figure 5.13, you will note that a DTE operating rate of 2.4, 4.8 or 9.6 kbps results in a line rate of 12.8 kbps, which is precisely one-fifth of the DS0 64 kbps data rate. Thus, multiplexers used by British Telecom at their local exchange are capable of placing five low-speed KiloStream circuits onto each DS0 channel.

The HDB3 line coding shown in Figure 5.13 represents the ITU method used to ensure an appropriate density on a T-carrier. You

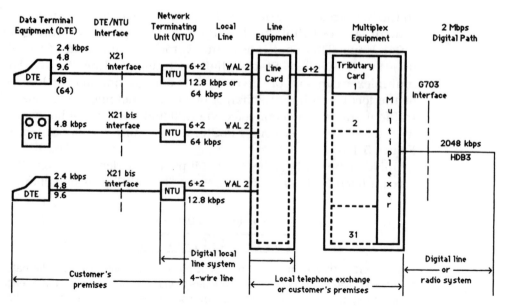

Figure 5.13 KiloStream structure

are referred to Chapter 6 for information concerning this method of zero suppression.

Representative multiplexers

Both British Telecom as well as several third-party vendors market a variety of multiplexers to extend the functionality and capability of KiloStream usage. Two representative products marketed by British Telecom that warrant discussion are the K3 and K5 multiplexers.

The British Telecom K3 multiplexer enables the KiloStream user to obtain a voice and data transmission capability on a 64 kbps channel. This multiplexer includes one 32 kbps CVSD modulation adapter that digitizes voice input at 32 kbps. In addition to voice digitization, the K3 also supports up to two synchronous channels at data rates up to 9.6 kbps on a 64 kbps KiloStream circuit.

The British Telecom K5 multiplexer can be used to maximize the data-carrying capacity of a 64 kbps KiloStream circuit. This multiplexer supports up to five synchronous channels at data rates up to 9.6 kbps or two channels at 19.2 kbps and one channel at 9.6 kbps for operation in a 64 kbps circuit.

Integrated Services Digital Network (ISDN)

Integrated Services Digital Network (ISDN) represents a communications carrier transmission facility developed as a mechanism to integrate the transportation of voice and data over a common facility. From the perspective of the subscriber or end-user, ISDN permits data to be transmitted end-to-end in a digital format. This means that the error rate on an ISDN connection should normally be lower than that obtainable on a modem call initiated over the PSTN, even though both calls could conceivably be routed over the same long-distance digital infrastructure. This is because the local loop ISDN transmission is digital, whereas the use of a modem would result in analog transmission occurring on the local loop. In this section we will first turn our attention to obtaining an appreciation of ISDN, including its two user interfaces, its network characteristics, terminal equipment, and the line signaling method used on the local loop. This will be followed by a brief discussion of the use of switched ISDN as an introduction to the succeeding section in this chapter which is focused on switched digital services, including ISDN.

ISDN user interfaces

There are two ISDN user interfaces—the Basic Rate Interface (BRI) and the Primary Rate Interface (PRI). Each interface consists of a group of 64 kbps channels referred to as B (Bearer) channels, and one signaling channel known as a Data (D) channel. The B channels are used individually to establish a voice or 64 kbps data call, or can be aggregated to place an $N \times 64$ kbps data call, where N refers to the number of aggregated B channels. The Data (D) channel is a packet-switched channel which is connected at the communications carrier's central office to a separate packet network which is shared by all ISDN users.

The Basic Rate Interface (BRI)

The ISDN Basic Rate Interface (BRI) consists of two 64 kbps B channels and one separate 16 kbps D channel. The BRI connection is formed by multiplexing the three channels by time on a twisted-pair wire routed between an end-user terminal device and a communications carrier central office or a local Private Branch Exchange (PBX). Figure 5.14 illustrates the ISDN Basic Rate Interface time division multiplexed channel format.

In examining Figure 5.14 note that 48 kbps is used for framing, resulting in the actual operating rate of the BRI being 192 kbps. However, only 144 kbps are used to transport voice and data.

The D channel was designed both for controlling the B channels through the sharing of network signaling functions on this channel as well as for the transmission of packet switched data. Signaling functions include call setup, call deflection in which a central office is informed to reroute calls to another specified number, and call

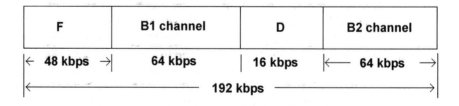

Figure 5.14 The ISDN Basic Rate Interface channel format

hold in which a user can place a caller on hold and either take or make a new call.

Concerning the transmission of packet switched data, the D channel provides the capability for a number of applications, including monitoring home alarm systems and the reading of utility meters upon demand. Since these types of applications have minimum data transmission requirements, the D channel can be used for a variety of applications in addition to providing the signaling required to set up calls on the B channels. In fact, one of the more recent ISDN innovations is an 'always on' D channel. This feature significantly reduces call setup time and can also be used to expediently pass aggregation requests so a user can group two B channels together to obtain a 128 kbps transmission capability.

The Primary Rate Interface (PRI)

The Primary Rate Interface (PRI) represents a multiplexing arrangement whereby a grouping of Basic Rate Interface users share a common line facility. Typically, Primary Rate Interface access will be employed to directly connect a Private Automated Branch Exchange (PABX) to the ISDN network. This access method is designed to eliminate the necessity of providing individual basic access lines when a group of terminal devices shares a common PABX which could be directly connected to an ISDN network via a single high speed line. Due to the different types of T1 and E1 network facilities in North America and Europe, two primary access standards have been developed.

In North America, primary access consists of a grouping of 23 B channels and one D channel to produce a 1.544 Mbps composite data rate, which is the standard T1-carrier data rate. In Europe, primary access consists of a grouping of 30 B channels plus one D channel to produce a 2.04 Mbps data rate, the E1-carrier transmission rate in Europe.

Functional groupings and reference points

One of the key elements of ISDN is a small set of compatible multipurpose user–network interfaces that were developed to support a wide range of applications. These network interfaces are based upon the concept of a series of reference points for different user terminal arrangements which is then used to define these interfaces. Figure 5.15 illustrates the relationship between ISDN reference points and network interfaces.

The ISDN reference configuration consists of functional groupings and reference points at which physical interfaces may exist. The functional groupings are sets of functions that may be required at an interface, while reference points are employed to divide the functional groups into distinct entities. To obtain an appreciation for the different functional groupings, let's examine those groupings in conjunction with the four major ISDN protocol reference points. In doing so, we will discuss the labels in each of the boxes and above the dashed lines in Figure 5.15.

The TE (terminal equipment) functional grouping is comprised of TE1 and TE2 type equipment. Examples of TE equipment include digital telephones, conventional data terminals, and integrated voice/data workstations.

TE1

TE1 type equipment complies with the ISDN user–network interface and permits such equipment to be directly connected to an ISDN 'S' type interface that supports multiple B and D channels. An example of a TE1 compatible device is a digital telephone. TE1 equipment connects to ISDN via a twisted-pair four-wire circuit. Transmission is full-duplex and occurs at 192 kbps for basic access and at 1.544 or 2.048 Mbps for primary access.

TE2

TE2 type equipment represents devices with non-ISDN interfaces, such as RS-232 or the ITU X or V Series interfaces. This type of

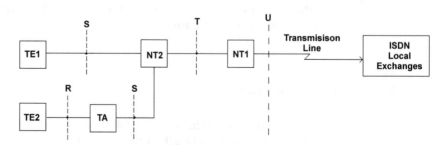

Figure 5.15 ISDN reference points and network interfaces

equipment must be connected through a TA (terminal adapter) functional grouping, which in effect converts a non-ISDN interface (R) into an ISDN Sending interface (S), performing both a physical interface conversion and a protocol conversion to permit a TE2 terminal to operate on ISDN.

Terminal adapters

Due to the large base of non-ISDN equipment currently in operation, the terminal adapter (TA) plays an important role if the use of this digital network expands. The terminal adapter performs a series of functions to convert non-ISDN equipment for use on the ISDN network. First, it must adapt the data rate of the non-ISDN device to either a 64 kbps B channel or a 16 kbps D channel operating rate. Next, it must perform the conversion of data from the non-ISDN device to a format acceptable to ISDN. For example, a non-ISDN device such as an intelligent modem might have its AT commands converted into ISDN D-channel signaling information.

Since a basic access channel operates at a multiple of most non-ISDN equipment rates, most terminal adapters include a multiple number of R interface ports. This allows, for example, an asynchronous modem connected to a personal computer, a facsimile machine and a telephone to be connected to a basic access line via the R interface. In fact, most commercially available adapters have three R interfaces.

Rate adaption

Rate adaption represents a technique which allows non-ISDN devices that operate at data rates below 64 kbps to communicate at 64 kbps. The first rate adaption standard, V.110, was approved by the ITU in 1984 for synchronously operated devices. In 1988 that standard was modified to support asynchronous transmission. Some of the problems associated with the V.110 standard include its failure to completely define flow control, lack of error detection, and difficulty of detecting and adjusting to a change in the data rate of non-ISDN devices. Due to those problems, a second rate adaption standard was originated in the United States by the ANSI T1/E1 standards committee. Known as V.120, this standard is based upon the HDLC protocol which makes flow control, error detection and correction and other functions very easy to perform.

Under the V.120 standard, rate adaption occurs through 'flag stuffing' in which a sufficient number of flag (01111110) bytes are added to a non-64 kbps data stream to increase its operating rate to 64 kbps.

Network terminators

There are two network terminators associated with ISDN, NT2 and NT1 (Figure 5.15). NT2 is used to interface optional customer switching equipment, such as a PBX. NT1 functions as a bridge between the local loop wiring and customer wiring, with a single NT1 capable of supporting multiple terminal adapters.

- The NT1 (network termination 1) functional group is the ISDN digital interface point and is equivalent to layer 1 of the OSI reference model. Functions of NT1 include the physical and electrical termination of the loop, line monitoring, timing and bit multiplexing. In Europe, where most communications carriers are government-owned monopolies, NT1 and NT2 functions may be combined into a common device, such as a PABX. In such situations, the equipment serves as an NT12 functional group. In comparison, in the United States the communications carrier may provide only the NT1, while third-party vendors can provide NT2 equipment. In such situations, the third-party equipment would connect to the communications carrier equipment at the T interface.

- The NT2 (network termination 2) functional group includes devices that perform switching and data concentration functions equivalent to the first three layers of the OSI Reference Model. Typical NT2 equipment can include PABXs, terminal controllers, concentrators and multiplexers.

Interfaces

In this section we will briefly review the functions associated with the ISDN R, S, T and U interfaces (Figure 5.15).

- As previously noted, the R interface is the point of connection between non-ISDN equipment and a terminal adapter. Although the R interface can consist of any common DTE interface, most terminal adapters support RS-232 and V.35.

- The S interface is the standard interface between the TE and NT1 or between the TA and NT1. This is the interface to a 192 kbps, 2B+D, four-wire circuit. The S/T interface can operate up to distances of 1000 m using a pseudo-ternary coding technique. In this coding technique, a binary 1 is encoded by the transmission of an electrical 0, while binary 0s are encoded by transmitting alternating positive and negative pulses. Since there are three signal states (+, 0, −voltage) used to encode two symbols, the coding method is called pseudo-ternary. An example of this coding technique is illustrated in Figure 5.16.

- The T interface is the customer end of an NT1 onto which you connect an NT2. For basic access, the T interface is a 192 kbps, four-wire, 2B+D interface. For primary access, the T interface is a 1.544 Mbps, 23B+D, four-wire circuit or a 2.048 Mbps, 30B+D, four-wire circuit.

- The U interface is the ISDN reference point that occurs between the NT1 and the network and is the first reference point at the customer premises. The coding scheme for information on the U interface is known as 2B1Q, an acronym for 'two binary, one quaternary'. Under this coding scheme every two bits are encoded into one of four distinct states that are known as quats. The left-hand part of Figure 5.17 illustrates an example of 2B1Q encoding, while the right-hand part of that illustration indicates the relationship between each dibit value and its quats code.

  From the U interface, transmission occurs at 160 kbps to the telephone company central office. Due to the use of 2B1Q coding, a maximum transmission distance of 18 000 feet (5500 m) is supported at a data rate of 160 kbps. This data rate represents 144 kbps used for the 2B+D channels and 16 kbps used for synchronization. In comparison, from the U

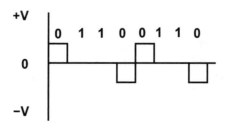

Figure 5.16 Pseudo-ternary coding example

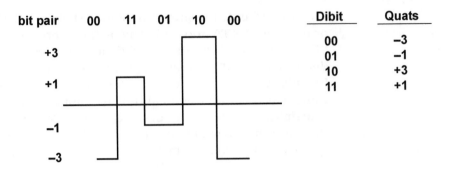

Figure 5.17   2B1Q coding example

interface to the S/T interface the data rate is 192 kbps, with 48 kbps used for synchronization and line balancing.

Switched network operation

ISDN represents a popular method for initiating switched 56/64 kbps calls over the digital network. Such calls can represent individual voice conversations that can be routed to a telephone connected to the analog PSTN or another ISDN terminal device, or aggregated to obtain an $N \times 56/64$ kbps transmission capacity.

When ISDN was in its infancy, calls were billed based upon duration even when such calls were local. More recently, many communications carriers did away with a per minute local use charge and now bill basic access at a fixed monthly fee. That fee, which can range between $30 and $50, while higher than normal telephone charges, enables subscribers to transfer data at up to 128 kbps by aggregating two B channels. In the commercial area, Primary Rate ISDN enables subscribers to aggregate up to 24 or 30 B channels to support high speed data transfer or videoconferencing on demand. When used to aggregate B-channel calls an inverse multiplexer that complies with what is known as the 'BONDING' standard is commonly used. The inverse multiplexer controls the dialing of multiple calls, either via the connection of individual Basic Rate Interface access lines or via a Primary Rate Interface line, aggregating the bandwidth of the composite B channels into a serial data stream used to connect high speed devices, such as videoconference equipment, at geographically separated areas.

Figure 5.18 illustrates the use of an inverse multiplexer to aggregate the use of 6 B channels on three ISDN Basic Rate Interface connections. Note that since the inverse multiplexer aggregates the use of 6 B channels, it supports videoconferencing at $6 \times 56/64$ kbps. Also note that the splitting of the composite data stream generated by the videoconference system is the inverse of multiplexing, resulting in the term 'inverse multiplexer' used for this device.

A second common use of ISDN is to provide a router connection for geographically separated locations or to connect a location's LAN to the Internet. For both types of router communications ISDN can be used in several ways. First, you can install a leased line ISDN connection which is billed on a monthly basis regardless of usage. Secondly, you can use dial-ISDN to obtain a temporary connection between two locations. A dial-ISDN connection can function either as an alternative to a leased line or to supplement a leased line. Concerning the former, if your organization requires only a few hours of communications between locations each day it may be less expensive to use dial-ISDN instead of a leased line connection. Concerning the use of dial-ISDN as a supplement to the use of a leased line, often line utilization may approach 100% only during a short period of time. Instead of obtaining a second leased line or upgrading an existing line to obtain an increase in bandwidth that would be billed on a monthly basis, it is often significantly less expensive to use dial-ISDN as a supplement to an existing leased line.

To support the automatic use of ISDN you could obtain a pair of inverse multiplexers which would provide an interface between the router and the leased line and one or more dial-ISDN lines. The inverse multiplexer would monitor the level of utilization on the leased line and initiate one or more dialed ISDN connections when the level of utilization on the leased line reached a predefined level.

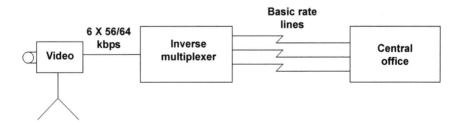

Figure 5.18 Using an inverse multiplexer to aggregate the capacity of three basic rate interface ISDN lines

Switched digital services

In the previous section in this chapter we examined the use of ISDN B channels as a mechanism to obtain a switched digital transmission capability. ISDN represents one of several methods that can be employed to initiate switched digital calls. In this section we will turn our attention to the different types of switched digital services, including ISDN.

Dataphone Digital Service (DDS)

The first commonly available method of switched digital transmission dates from AT&T's introduction of Dataphone Digital Service (DDS). Under DDS AT&T offered a switched 56 kbps digital transmission capability.

Figure 5.19 illustrates one method by which DDS subscribers obtain switched 56 kbps access. Note that a leased line is required to connect the subscriber's location to the serving carrier's central office. At that office the leased line is connected at a Point-of-Presence (POP) to the selected interexchange carrier. Although AT&T was the first vendor to offer switched 56 kbps digital service, it is now available from other interexchange carriers as well as from the Regional Bell Operating Companies (RBOCs).

In addition to the use of a 56 kbps leased line to the local central office, organizations can access switched 56 kbps using three additional methods. Those methods include the use of Digital Switched Access (DSA), T1 access, or a connection through the use of an ISDN Basic Rate Interface. Figure 5.20 illustrates the use of each of those three methods.

Digital Switched Access (DSA)

DSA (Figure 5.20A) can be ordered from both local telephone companies as well as interexchange carriers. This method of

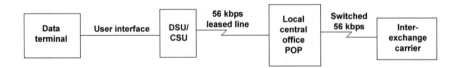

Figure 5.19 Switched 56 kbps Dataphone Digital Service access

switched 56 kbps access is simply a digital connection from the subscriber's premises to the local ESS 5 switch, which is then connected to an interexchange carrier's central office. When available, its monthly cost is typically a fraction of the cost of the switched DDS access charge.

Time slot switched access

When using a T1 connection (Figure 5.20B) one or more DS0 time slot 64 kbps channels can be dedicated to switched 56 kbps service. Instead of a data terminal connected to a DSU/CSU you would use a T1 access device, such as a T1 multiplexer, PBX or channel bank as a connection point for your data terminal. Since the multiplexer, PBX or channel bank includes a built-in

A. Digital Switched Access

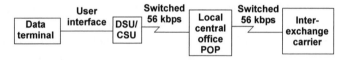

B. T1 Access

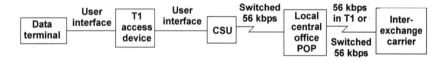

C. ISDN BRI

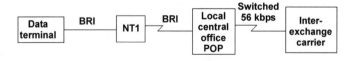

Legend:
POP Point-of-Presence
BRI   Basic Rate Interface

Figure 5.20 Alternative switched 56 kbps access methods

DSU, the access device would then be connected to a stand-alone CSU.

ISDN BRI

Figure 5.20C shows the use of an ISDN BRI connection to obtain access to switched 56 kbps communications. This connection requires end-user equipment to perform 64 kbps/56 kbps rate adaption. A fourth method, which is not illustrated in Figure 5.20, represents a combination of T1 and ISDN BRI access. That method involves the use of one or more 64 kbps channels on an ISDN PRI interface.

Signaling incompatibilities

The original design of the T-carrier's time slots resulted in signaling information being carried 'in-band' through a process referred to as 'bit robbing'. This process, which will be illustrated in detail later in this book, requires the use of one bit in each byte in the sixth and twelfth frames of the T1 D4 framing format. As a result of the bit-robbing process the use of a 64 kbps channel was limited to a 56 kbps data transport capability. In comparison, ISDN uses a separate 'out-of-band' signaling network, referred to as Signaling System 7 (SS7), to transport call setup information, whose use frees up the data network so that the full 64 kbps capacity of a channel can be used. This technique results in what is referred to as a 'clear channel' or '64 kbps clear channel' transmission capability.

Although ISDN was available in approximately 70% of the United States and in a correspondingly large percentage of locations in Western Europe, it was far from universal in coverage when this book was written. In addition, as an ISDN call is routed between interoffice networks, an office that does not support ISDN will limit its data transport capacity to 56 kbps per channel. This explains why some ISDN calls may be completed at 56 kbps even though two locations communicating with each other both have switched digital access lines that support 64 kbps transmission.

The digital dialing plan

In the United States customers who subscribe to a Local Exchange Carrier's (LEC) Digital Switched Access facility are assigned a unique 10-digit number from the LEC. For both

56 kbps DSA and 64 kbps BRI customers, the 10-digit number has the format

NPA-NXX-XXXX

where NPA represents a three-digit area code, N represents any number from 2 to 9, and X represents any number from 0 through 9.

Dedicated access arrangements such as the use of DDS or a channel on a T1 use a '700 number' dialing plan. The format of those numbers currently is

700-56X-XXXX
or 700-737-XXXX

Table 5.4 indicates the dialing plans used for 56/64 kbps switched digital calls for inbound and outbound calling based upon access obtained by DDS, DSA and ISDN PRI.

To illustrate the use of the dialing plan information contained in Table 5.4, assume users in the United States and France wish to communicate with one another. To reach an ISDN subscriber in France an American would use one of three dialing formats based upon the type of switched digital access method. Those dialing methods include

1 73 33 (F) XX XX XX XX (DDS/T1)
0 11 33 (F) XX XX XX XX (DSA)
9 011 33 (F) XX XX XX XX (ISDN PRI 64 kbps)

where F stands for 1 for calls to the Paris region.

Reversing the calling direction, assume an ISDN subscriber in France wishes to communicate with a digital switched network

Table 5.4 Switched 56/64 kbps dialling plans.

Access method	Calling from	Calling to	Dialling plan*
DDS via T1 (56 kbps)	United States	Foreign country	173 + CC + NN-XXX-XXXX
DSA	United States	Foreign country	011 + CC + NN-XXX-XXXX
PRI ISDN	United States	Foreign country	011 + CC + NN-XXX-XXXX
DDS via T1 (56 kbps)	Foreign country	United States	1-700-56X-XXXX
DSA	Foreign country	United States	1-NPA-NXX-XXXX
PRI ISDN	Foreign country	United States	1-700-737-XXXX

*CC = country code, N = any number from 2 to 9, X = any number from 0 to 9.

user in the United States. The French ISDN user would use one of the following international dialing formats based upon the American's method of switched digital access

19 09 1 700 56X XXXX (DDS)
19 09 1 NPA NXX XXXX (DSA/BRI)
19 09 1 700 737 XXXX (ISDN PRI 64 kbps)

Broadband switched access

Included in the series of ISDN standards are a series of H channel specifications that were designed to transport user information at data rates well in excess of 64 kbps. Referred to as broadband switched access, data rates from 384 kbps are supported on different types of H channels. Table 5.5 lists presently defined ISDN H channel operating rates.

Several communications carriers, including MCI and AT&T, currently offer ISDN H0 and H11 services. ISDN H0 provides subscribers with 384 kbps of contiguous bandwidth which is the equivalent of six B channels. H0 service results in the allocation of 384 kbps through an ISDN network as a single data call.

Access to H0 service is commonly obtained by the use of a T1 line from a subscriber's premises to the serving central office. Although only six 64 kbps channels are used on the T1 access line, customers are billed for a full T1. Thus, many organizations commonly use a T1 access line to provide access to a variety of voice and data services to economize upon the monthly recurring cost of the line.

A second type of switched broadband ISDN service is ISDN H11. ISDN H11 provides a single 1.536 Mbps channel which is equivalent to 24 64-kbps channels. Access to ISDN H11 is accomplished by the use of a T1 circuit, with 8 kbps of the

Table 5.5 ISDN H channel operating rates.

H channel	Operating rate
H0	384 kbps
H11	1.536 Mbps
H12	1.920 MBPS
H21	32.768 Mbps
H22	44.160 Mbps
H4	135.168 Mbps

1.544 Mbps operating rate of the circuit used for framing, resulting in an operating rate of 1.536 Mbps.

To illustrate the economics associated with the use of switched digital service, I contacted AT&T for tariff information. Table 5.6 indicates the approximate day rate for several types of switched digital calls between New York and Los Angeles. Readers should note that AT&T and other communications carriers offer a variety of customized rates and frequently modify their charges.

Now that we have an appreciation for switched digital services, let's turn our attention to the method by which many of those services are accessed. That method is through the use of a T-carrier in North America and an E-carrier in Europe.

The T- and E-carriers

Due to the superior quality provided by T- and E-carrier facilities for supporting voice and data transmission, they are now routinely available for commercial use in most parts of North America, Europe and Japan. In this section we will examine the physical construction of the T- and E-carrier facility from the carrier's serving office to an organization's premises. In addition, due to the divestiture by AT&T of its operating companies, we will also focus our attention upon the pricing components related to the installation and operation of a T1 circuit that crosses local access transport area (LATA) boundaries in the United States. Using the preceding information as a base, we will then examine how these high-speed transmission facilities are typically used by end-user organizations.

The T- and E-carrier line structure

Unlike the interface for DDS that requires the use of a combined DSU/CSU or separate devices, only a CSU is required to terminate a T-carrier in North America and an E-carrier in

Table 5.6 AT&T switched digital tariffs for a Los Angeles to New York daytime call.

Service	Cost per minute
56/64 BRI to BRI	$0.35
56/64 PRI to PRI	$0.20
Switched 384 kbps	$0.70
Switched 1.536 Mbps	$1.60

Europe. The removal of the DSU results from the incorporation of its functionality into T1 and E1 multiplexers and channel banks.

Figure 5.21 illustrates the T1/E1 line structure routing from a communications carrier's nearest serving office to the customer premises. The first repeater located outside the serving central office is typically located 4500 feet (1370 m) from that office. Thereafter, T1 or E1 repeaters are typically spaced 6000 feet (1830 m) apart from one another.

Since repeaters are active devices that regenerate digital pulses, they require power. Power for repeaters is provided by the communications carrier which uses a power supply at the serving central office to place a voltage between the tip and ring leads on the circuit. Both the voltage and current vary depending upon the length of the line to the customer premises and the gauge of the wire. For example, the most common voltage used is a nominal −48 volt DC source, resulting in a −60 mA current on a low power T1 circuit and a −140 mA current on a standard T1 circuit. This type of powered circuit, which is completed by the CSU, is commonly referred to as a 'Wet T1' or 'Wet E1'. A second type of T1/E1 line provides a data transportation path through the use of optical fiber that transports light energy instead of electrical energy. When an optical fiber line is used to connect the customer premises to the serving central office, the facility is referred to as a 'Dry T1' or 'Dry E1' line due to the absence of power.

Since T1 and E1 CSUs were originally constructed to obtain power from the line, equipment that is not capable of being locally powered cannot be used with a dry T1 or dry E1 line.

T1 cost components

Prior to 1984 when AT&Ts divestiture became effective, there were only two types of tariffs — interstate and intrastate. Interstate tariffs were filed by AT&T and other long-distance carriers with the

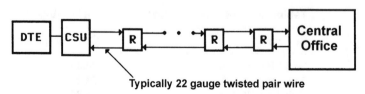

Typically 22 gauge twisted pair wire

Figure 5.21 T1/E1 line structure: CSU = channel service unit, R = repeater, DTE = data terminal equipment

Federal Communications Commission, while intrastate tariffs were filed by local operating companies with state public utility commissions.

Although the distinction between interstate and intrastate communications is still used to determine the applicable regulatory body, a new criterion with respect to line costs resulted from divestiture. This criterion concerns whether or not a service is within the local area served by the divested local operating companies. These areas are known as local access and transport areas, or LATAs, and approximately correspond to the standard metropolitan statistical areas defined by the US Commerce Department.

If service provided by a communications carrier links two locations within a LATA, it is an inter-LATA service. If a service must cross LATA boundaries, the local exchange carrier (LEC) must connect their facility to an inter-LATA carrier, also known as an interexchange carrier (IXC). Both IXCs and LECs have tariffs. These tariffs govern the use of IXC facilities and the use of local facilities by the IXC, the latter being the method by which LECs charge IXCs for the use of their local access facilities.

Based upon the preceding, end-users may have to contend with up to six types of tariffs: inter-LATA, intra-LATA and LATA access, for both interstate and intrastate communications. Within each LATA is a series of interface points known as points-of-presence (POP). Each IXC, such as AT&T, MCI and Sprint, has its own POPs which are the only locations within a LATA where the IXC can receive and deliver traffic. Due to this structure, a customer's premises must use the facilities of the LEC to connect to the IXC's network at the latter's POP.

In our examination of the cost elements of a T1 facility we will focus our attention upon an inter-LATA circuit. This type of circuit will consist of three components:

- a local circuit from the end-user's near-end premises to the POP of their selected IXC,

- an interoffice circuit that connects the near-end IXC's POP to the far-end IXC's POP,

- a local circuit from the IXC's far-end POP to the end-user's far-end premises.

Figure 5.22 illustrates an example of the routing structure of a T1 circuit between Macon, Georgia, and Washington, DC. Note that the local access channel in Macon, Georgia, is normally provided

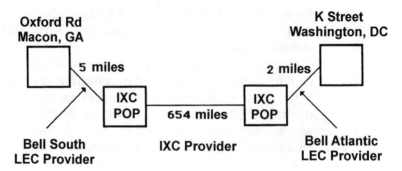

Figure 5.22 Routing structure of a T1 circuit: LEC = local exchange carrier, IXC = interexchange carrier, POP = point-of-presence

by Bell South, while the local access channel in Washington, DC. would be provided by Bell Atlantic. To minimize the cost associated with local access channels, many organizations are using the facilities offered by cable TV (CATV), fiber optic local transmission systems and microwave. Collectively, these techniques are known as bypass access since they bypass the LEC and route the user's organization to the IXC point-of-presence.

Although access to the IXC normally occurs through the use of LEC facilities, many organizations also employ bypass facilities to access the IXC point-of-presence. In many instances, the IXC point-of-presence may be in the same building occupied by the LEC. Thus, the connection between the two may be no more than a short cable connecting the equipment of the two communication carriers. Although most IXCs have a POP in each LATA, you should note that this does not ensure that all IXC facilities are available for connection. For example, some AT&T POPs may not support DDS or T1 connections.

For illustrative purposes, Table 5.7 lists the recurring (monthly) and one-time (installation) costs associated with the T1 circuit shown in Figure 5.22. The entries in this table are approximations used for illustrative purposes and, while they were relatively accurate at the time this book was published, you should note that they may have been rendered obsolete due to tariff changes.

The local access channel costs, in many instances, are very disproportionate to the circuit mileage. Due to this, many end-user organizations have employed the use of third-party carriers to obtain access to the IXC POP via coaxial cable, microwave or fiber optic transmission. As previously discussed, this method of accessing the POP is known as bypass access.

Table 5.7 T1 circuit cost components.

	Monthly ($)	Installation ($)
Local access channel		
Macon, GA		
Fixed	391	1000
Mileage 5 at $40	200	
Central office connection	62	110
Access coordination	22	207
Interchange channel		
Fixed	500	
Mileage 650 at $4	2600	
Local access channel		
Washington, DC		
Fixed	500	375
Mileage 2 at $45	90	
Central office location	62	110
Access coordination	22	207
Total	4449	2009

The central office connection and access coordination fees are normally billed by the IXC. The central office connection fee is the cost for the IXC connecting the local channel to an interoffice channel. The access coordination fee covers the cost of the IXC providing a single point of contact by coordinating installation and testing with the LEC.

Digital access and cross connect

In 1981 AT&T introduced its digital access and cross connect system (DACS) to facilitate testing as well as to reduce maintenance costs. Through the use of DACS, DS0 time slots can be dropped from one T1 circuit and added to another T1 line, a process known as drop and insert.

DACS cross connections are administered via software control. Originally this control was restricted to initiation from a communications carrier central office. Later, the ability to control DACS cross connections was extended to end-user locations. The latter capability is known as customer controlled reconfiguration (CCR) and is a service function of AT&Ts Accunet™ 1.5 service, as well as a function available from other communications carriers.

DACS have the capability to switch both subrate and full rate T1 channels. For internal networks, one vendor now markets an IBM

PC compatible personal computer with two add-in boards that can terminate four T1 circuits and switch all of the DS0 subchannels to or from each T1 line. With another vendor's PC product, internal corporate network users can terminate up to 16 T1 lines and manage up to 384 individual DS0 subchannels. Although the preceding general discussion of DACS was oriented toward T1 lines, a similar capability is associated with E1 lines. No matter where performed or the type of computer control used, DACS perform certain basic functions to include drop/insert/bypass and groom and fill operations.

Drop/insert/bypass

The drop/insert/bypass capability is used to satisfy the requirement for some voice and data channels to terminate at a node while other channels must bypass a node and are then routed on another circuit towards another node.

When a channel is dropped at a node or bypasses a node and is routed onto a different circuit, its drop or removal results in wasted capacity. This capacity can be filled by inserting other channels that have the same intermediate or final destination as the channel that bypassed a node. Figure 5.23A illustrates a DACS bypass operation.

Groom and fill operations

In addition to drop/insert/bypass, some DACS perform a groom and fill operation on rerouted data to maximize the data handling efficiency of the facility. The groom function is used to segregate channels for transmission to their appropriate end points as illustrated in Figure 5.23B. The fill function is often used with the groom function as the former maximizes the efficiency of a large number of groom operations. Here the fill feature attempts to combine traffic from two or more carriers onto one carrier with the same route destination. This feature, which is illustrated in Figure 5.23C, as well as the groom function, is usually incorporated into large DACS systems that control carrier facilities.

One example of the use of DACS within a carrier's communications network is shown in Figure 5.24. which illustrates British Telecom's KiloStream network routing. In this example, note that switching occurs on 64 kbps DS0 channels, which explains why low-speed 2.4 through 9.6 kbps data streams are

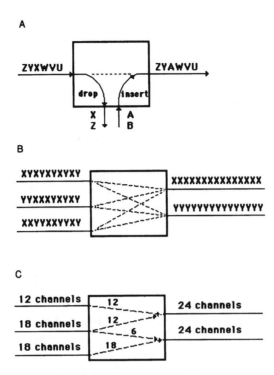

Figure 5.23 DACS functions: A drop/insert/bypass; B groom operation; C fill operation

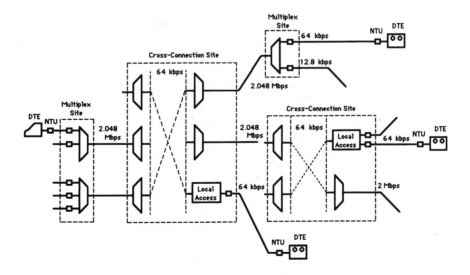

Figure 5.24 KiloStream network routing

operated at a line speed of 12.8 kbps on KiloStream, since five of those data streams can then be carried within a switched DS0 channel.

Accunet T1.5 and MegaStream utilization

Both Accunet T1.5 offered by AT&T and MegaStream offered by British Telecom are T- and E-carrier transmission facilities for high speed data and high volume speech communications.

The typical utilization of each T- and E-carrier is a transportation facility for data sources that can be effectively serviced by the high data rate provided by the carrier. Such usage includes the servicing of digital PABXs, analog PABXs with voice digitizers, clustered terminals via multiplexer input analog terminations, data circuits, computer-to-computer transmission and videoconferencing.

The Accunet T1.5 facility operates at 1.544 Mbps, while the MegaStream facility operates at the CEPT 2.048 Mbps data rate.

The typical utilization of an Accunet T1.5/MegaStream facility is illustrated in Figure 5.25.

The key to the effective use of T- and E-carrier facilities is the selection of an appropriate multiplexer. In Chapter 8 we will focus

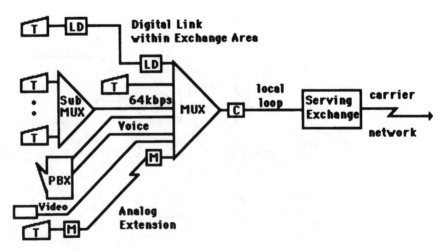

Figure 5.25 Typical T- and E-carrier utilization. Through the use of T- and E-carrier facilities a mixture of voice, data and video transmission can be routed over a common high speed circuit: LD = line driver, T = terminal, M = modem, C = channel service unit, MUX = multiplexer

our attention upon the operational capabilities of this category of communications equipment.

Fractional T1 and E1

Until 1989, organizations that required a data transportation capability in excess of that offered on DDS or KiloStream were forced to either migrate to a T- or E-carrier or obtain multiple subrate circuits. In migrating to a T- or E-carrier the end-user had to pay for 24 or 30 64-kbps channels on a four-wire circuit regardless of the number of channels they actually needed. If multiple subrate circuits were used, the cost associated with the central office connection and coordination fees, as well as multiple DTEs, such as multiplexers, usually made that solution uneconomical.

Based upon the preceding, AT&T was one of several communications carriers to introduce a fractional T1 service. Marketed under the name Accunet Spectrum of Digital Services (ASDS), users of this service can lease a fractional portion of the T1 bandwidth in 64 kbps DS0 increments. ASDS provides digital interoffice channels at 56 or 64 kbps and intermediate bit rates of 128, 256, 384 and 768 kbps. Other communications carriers, including MCI, Sprint and the RBOCs, as well as several European communications carriers, offer fractional T1 and E1 services.

The benefits of fractional T1 and E1 are considerable. In addition to saving the cost of leasing an entire T1 or E1 line when only a portion of the bandwidth is required, this facility also provides a method for orderly growth. Thus, an organization could grow from 128 kbps today to 256 kbps tomorrow. Another significant advantage of fractional T1 and fractional E1 concerns cost. On an IXC basis the cost of an FT1 or FE1 transmission facility is significantly less than the cost of a full T1 or full E1 circuit. Thus, instead of having to pay for a full T1 or E1 circuit while only requiring a small portion of its capacity, users now have the ability to select an increment of bandwidth more appropriate to their transmission requirements.

Access to FT1 service is obtained by the installation of a T1 line from the serving central office to the subscriber. When that access line is established via the use of copper cable it must carry no more than 15 zeros in a row to provide proper timing for repeaters on the span line and as a self-clocking mechanism for the receiver. Several types of line coding schemes can be selected to obtain an appropriate 1s density. One popular technique is Bipolar 8-Zero Substitution (B8ZS) in which a byte of zeros is converted into an intentional series of bipolar violations which are interpreted at the receiver as a sequence of eight zeros, enabling the contents of the byte to be restored while generating 1s which maintain timing. Chapter 6 provides a detailed examination of B8ZS coding.

Although B8ZS coding is commonly used to maintain a 1s density, its support requires carrier equipment compatibility. As an interim measure to provide a 1s density, some carriers support Alternate Channel Alternate Mark Inversion (ACAMI). Under ACAMI line coding, user data is assigned to alternate DS0s on the T1, starting from the lowest DS0 (i.e., DS0 1, 3, 5, ...). Each DS0 in between (DS0 2, 4, 6) is then populated with a fixed byte, with a 0 followed by seven 1s (01111111) recommended. ACAMI is not as efficient as B8ZS and restricts FT1 access to 784 kbps while preventing half of a T1 access line from being used. Thus, it should be used only when you require a clear channel and B8ZS coding is not available from an LEC.

In the United States access to FT1 is similar to the method illustrated in Figure 5.22 for access to a T1 circuit. At the time this book was written, all LECs provided access to FT1 by using a full T1 circuit that is interconnected at the POP at fractional T1 rates. In Europe a full E1 circuit routed from the subscriber's premises to the serving central office provides access to FE1 service.

T3 and FT3 operation and utilization

Although fractional T1 satisfies the user who does not require a full T1 circuit, it did not alleviate the problem large organizations faced when their data and voice transmission requirement began to occupy multiple T1 circuits. To provide assistance to large users in meeting their demand for higher capacity transmission, most communications carriers have made their DS3 facilities available for commercial use.

The DS3 signal operates at a rate of 44.736 Mbps and carries 28 T1 signals, which is the equivalent of 672 voice channels. The actual circuit used to transport a DS3 signal is referred to as a T3 transmission facility. Although many large organizations are using T3 transmission facilities to support access from the Internet to their World Wide Web servers, it may be quite some time before most organizations have to consider the use of this facility; its commercial availability provides a migration path that should satisfy the voice, data and video transmission requirements of a majority of organizations and governmental agencies. In the same way that many organizations require only a portion of a T1 line, other organizations require only part of the capacity of a T3 line. This resulted in many communications carriers introducing fractional T3 (FT3) transmission facilities.

REVIEW QUESTIONS

1 Name four types of digital transmission services you could use to access a Frame Relay network service.

2 What equipment would an organization use to transmit several asynchronous data sources over one synchronous DDS network facility?

3 What is the purpose of the control (C) bit in a DDS byte?

4 Compare the efficiency of DDS byte stuffing to DDS multiplexing. Which is more efficient?

5 Illustrate how a string of seven consecutive zeros would be encoded on DDS at a transmission rate of 56 kbps if the last binary 1 was encoded as a positive pulse.

6 Illustrate the encoded sequence a DSU/CSU operating at 56 kbps would generate when the loop-back button is pressed when the device operates at 56 kbps.

7 What is the diagnostic channel rate of a 2.4 kbps DDS II transmission facility?

8 Explain how you could use a 56 or 64 kbps DDS or KiloStream facility to carry voice and data.

9 Describe several examples of the potential use of the 2B+D channels that make up the ISDN Basic Rate Interface.

10 How is it possible for an ISDN BRI subscriber to obtain a 128 kbps data transmission capability.

11 What is the composition of an ISDN Primary Rate Interface line in North America and in Europe?

12 What is the purpose of ISDN reference points?

13 What is the purpose of an ISDN terminal adapter?

14 What is the purpose of rate adaption?

15 Discuss the reason why an ISDN call initialized using a 64 kbps B channel may result in a 56 kbps data transmission rate.

16 What type of transmission facility is used to obtain access to a switched 384 kbps transmission facility?

17 What is the difference between a 'wet' and 'dry' T1 line?

18 What is a point-of-presence?

19 What is the primary rationale behind bypassing a local exchange carrier to access the point-of-presence of an inter-exchange carrier?

20 How could you use a digital access cross connect system at location B to route three DS0 channels from location A to C via location B if T1 lines connect A to B and B to C?

T- AND E-CARRIER FRAMING AND CODING FORMATS

In this chapter we will focus our attention upon both North American and European T- and E-carrier framing and coding formats. Knowledge of framing and coding formats is extremely important as it provides a foundation for understanding how messages and error performance data can be conveyed over different types of T- and E-carrier transmission facilities. In addition, a solid understanding of the information presented in this chapter provides a foundation for understanding testing and troubleshooting methods described later in this book.

Our examination of T- and E-carrier framing will include an investigation of the different types of framing used, as well as the method by which mechanical signaling and performance information is transported, and how the frame structure is designed to provide synchronization and report alarm conditions. Since maintaining a minimum number of binary 1s on T- and E-carriers transported over copper cable is critical for repeater operations, we will also examine several coding formats that are employed to ensure that a minimum 1s density is provided on the transmission facility.

The examination of T- and E-carrier framing and coding formats will begin with the T1 and E1 signaling systems. Once this is accomplished we will then turn our attention to the T3 signaling system and the manner by which 28 T1 signals are multiplexed and transmitted at an operating rate of 44.736 Mbps.

6.1 T1 AND E1-CARRIER FRAMING

Since the framing structures used on North American and European T1- and E1-carrier facilities vary considerably from one another, we will examine each structure as a separate entity in this section.

North America

In North America the T1-carrier was designed to transmit 24 independent voice channels, with each channel encoded as a 64 kbps data stream.

Although we will use the terms T- and E-carrier to discuss different framing formats used on those carriers, in actuality those terms refer to transmission facilities. The actual frame formats used when transmitting data or digitized voice signals are more precisely referred to as Digital Signal (DS) levels. Recognizing the fact that these terms are commonly used interchangeably, we will also do so.

Signal synchronization

The development of channel banks to transport digitized voice resulted in the need to synchronize the data flow between those devices. Since the first series of channel banks were employed on point-to-point circuits, a loop timing method was selected as a mechanism to provide timing. Under loop timing a receiver takes its timing from the input data stream and uses that timing on its transmit side as a clocking mechanism, hence the term loop timing.

Figure 6.1 illustrates an example of loop timing which is commonly used for point-to-point asynchronous T-carrier systems. Loop timing, while easy to implement, requires a recovery mechanism if a receiver end of a transmission system drifts out

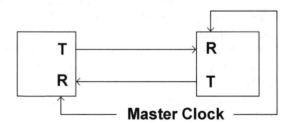

Figure 6.1 T- and E-carrier transmission facilities are asynchronous systems that obtain clocking via loop timing. Under loop timing received data provides clocking for the transmit side

of sequence with the transmit end. That recovery method is provided by the use of a framing bit added to each multiplexed frame transmitted between channel banks. Originally, the framing bit was developed for synchronization and clock recovery. However, as we will note in this section, modern framing is also used to transport messages as well as link performance data.

D1 framing

When the first digital channel bank was developed in 1962 a framing method was introduced which, in reflection, represents a simplistic method used to provide synchronization between transmitter and receiver. To understand that framing method, let's first review the manner by which a DS1 signal is formed.

In North America the DS1 signal represents a composite of 24 separate DS0 channels, each containing one PCM encoded voice signal digitized at 64 kbps. The 24 channels in a DS1 signal are multiplexed in a round-robin order to ensure each channel is transmitted in turn and that every channel receives a turn prior to any channel receiving a second turn. The sequence of multiplexing results in one eight-bit sample from each channel grouped in sequence to form a multiplexing frame. To denote the end of each frame and the beginning of the next frame, a special bit called the frame bit was added to the end of each frame. Since a sample from each DS0 channel is encoded into a PCM word using eight data bits, 24 DS0 channels represent a sequence of 192 data bits. The full pattern of one frame bit and 192 data bits is known as the DS1 (digital signal, level 1) frame and represents a total of 193 bits. Figure 6.2 illustrates the framing method used with D1 channel banks. Since sampling occurs 8000 times per second, 193 bits×8000 samples per second results in the 1.544 Mbps operating rate of a T1 circuit. As this operating rate includes 8000 frame bits, only the remaining 1.536 Mbps is actually available to the user. Now that we have an appreciation for the manner by which a DS1 signal is formed, let's turn our attention to D1 framing.

D1 framing was designed to enable synchronization to be maintained in the event of a single transmission error. In addition, it was designed to provide a rapid and low cost to implement resynchronization capability. To accomplish the preceding, the framing bit simply alternates between a pulse (1) and a space (0).

In addition to using a rather simplistic framing method, D1 channel banks used only seven bits in each eight-bit byte for data. The least significant bit in the byte is used for two-state signaling,

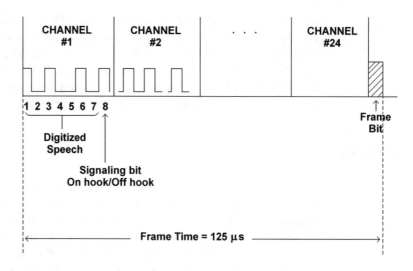

Frame bit pattern 1, 0, 1, 0, . . .

Figure 6.2 The framing method used with D1 channel banks that form a DS1 signal

permitting on- and off-hook information to be conveyed. If revertive pulse signaling was used, the pulses were transmitted in bit position 7, reducing the number of data bits to six. This resulted in D1 channel banks not having an adequate signal-to-noise ratio and resulted in the introduction of a series of modified channel banks. One such channel bank was the D2 which was introduced in 1969.

D4 framing

The introduction of the D2 channel bank resulted in several improvements to the quality of transported digitized voice signals as well as in the manner by which the DS1 signal was framed. D2 channel banks were designed so they were capable of converting 96 voice channels onto four independent 24-channel DS1 signals. To do so the D2 channel bank employed a single coder and compander for all 96 channels. The channel bank organized the 96 individual channels into eight groups of 12, with each channel in a group sampled sequentially. This was followed by interleaving the signals from each group to form a 96-channel signal used to create four DS1 signals. Accompanying the new grouping of 12 channels was a new framing method which enables each group of 12 channels to be identified. This method of framing is referred to as the Superframe Format and is also known as D4 framing. The term D4 dates to 1976 when AT&T introduced its D4 channel bank which uses the Superframe Format and μ-law companding with digitized speech samples assigned to a full eight-bit byte. However, as we will note later in this chapter, both D2 and D4 channel banks use a bit robbing procedure described in detail later in this section to support the transmission of signaling information. This action results in one bit in each byte in frames 6 and 12 being used to convey signaling information. This also explains why older digital transmission systems provide only a 56 kbps data transfer capability on 64 kbps DS0 channel while ISDN, which uses a separate signaling network, permits the full use of the 64 kbps DS0 bandwidth for data transmission.

Under D4 framing the frame bits in 12 consecutive frames are grouped together to form a superframe whose frame bits are used to form a repeating pattern. Figure 6.3 illustrates the D4 framing structure and framing pattern. Under D4 framing, the 1.544 Mbps data stream must meet the following requirements:

- It must be encoded as a bipolar, AMI, non-return to zero signal to ensure that the signal has no dc component and can be transformer coupled, permitting the circuit to carry power for the repeaters.

- Each pulse must have a 50% duty cycle with a nominal voltage of 3.0 volts.

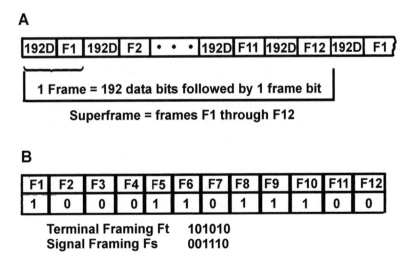

Figure 6.3 D4 framing structure and framing pattern. The D4 framing pattern represents the hex characters 8CD which are continuously repeated: A D4 framing structure; B D4 framing pattern

- There can be no more than 15 consecutive 0s present in the data stream, which defines the minimum 1s density of the circuit.
- The D4 framing pattern is embedded in the data stream.

The framing bits in the D4 superframe consist of six Ft (terminal framing) bits that are used to synchronize the bit stream and six Fs (signal framing) bits that are used to define multiframe boundaries as well as to identify what is known as robbed bit signaling in frames 6 and 12. Robbed bit signaling is discussed later in this chapter.

The Ft bit conveys a pattern of alternating 1s and 0s (101010), which is employed as a mechanism to synchronize terminal equipment. The Ft bit sequence is also used to define frame boundaries, enabling one slot to be distinguished from another. Due to this, it is also known as a frame alignment signal. The Fs bit conveys a pattern of 001110, which is used to define multiframe boundaries. This enables one frame to be distinguished from another, permitting frames 6 and 12 to be identified for the extraction of their signaling bits. Because the Fs bit sequence identifies frames containing signaling bits, its sequence is also referred to as signaling framing bits. Note that the composite D4 framing pattern represents the hex characters 8CD which are continuously repeated.

Since D4 framing represents a constant framing pattern, it can be utilized to determine the approximate bit error rate of a T1 line by monitoring. This can be accomplished by the use of a test set that can monitor the framing bit error rate which will usually reflect the error rate on the line. Unfortunately, for a more accurate measurement of the line error rate, transmission must be interrupted to permit the use of more sophisticated testing equipment.

Although D4 framing can be considered to represent a considerable improvement over the simplistic D1 framing sequence pattern of '1010...', it does not provide any performance information concerning the quality of transmission. In addition, for precise status information the entire T1 line has to be relinquished by the customer. Due to these problems a more modern framing method referred to as Extended Super Frame (ESF) was developed. Since both D4 and ESF support robbed bit signaling, let's first turn our attention to the manner by which bit robbing occurs when D4 framing is used. This will facilitate an explanation of bit robbing when ESF framing is used.

Bit robbing

In voice transmissions, mechanical signaling information, such as 'on-hook' and 'off-hook' conditions, dialing digits and call progress information, must be transmitted separately for each voice channel by including signaling information in the data stream. Until the development of a separate network to transport signaling information, this is accomplished by robbing the eighth bit in every sixth and twelfth frame to transmit and receive voice channel signaling information.

The process of transmitting signaling data associated with each voice circuit within the voice channel is known as associative signaling. In comparison, the use of a common channel dedicated to carrying the signaling data for all voice circuits within a T-carrier link is called common channel signaling.

Under the D4 format, channels 1–5 and 7–11 use all bits for information coding. Since there are no special bits assigned for signaling, the least significant bit position of each channel in every sixth and twelfth frame is robbed for signaling.

Bits taken from the sixth frame are referred to as 'A' bits, while bits taken from the twelfth frame are referred to as 'B' bits.

Figure 6.4 illustrates the bit robbing process used to convey voice signaling information. Note that the use of the least significant bit in each DS0 PCM word in the sixth and twelfth frames minimizes the duration of bit robbing and its effect upon a voice conversation. This is because each DS0 PCM word has a duration of 125 μs, resulting in 625 μs of full eight-bit PCM words (125×5) for every 125 μs, where the word loses one bit of accuracy.

Since only frames 6 and 12 contain associated signaling, those frames must be distinguishable from one another. This is the basis for the multiframe D4 structure consisting of 12 frames that are distinguished by the D4 framing pattern previously illustrated in Figure 6.3B.

One example of the use of robbed bit signaling is to pass E&M (ear and mouth) status information. Here the terms ear and mouth relate

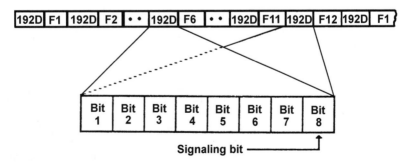

Figure 6.4 Bit robbing. Every eighth bit in the sixth and twelfth frames is used for signaling

to the speaker and receiver in a telephone handset. Under robbed bit signaling, E&M status information is passed by varying the setting of the A and B bits. When the A bit is set to 0 it denotes an active line, while it is set to 1 when the line is inactive. The B bit is used to pass on- and off-hook information, with the B bit set to 0 indicate an off-hook condition, while it is set to 1 to indicate an on-hook condition.

Extended superframe format

As previously discussed, D4 framing provides only an indirect measurement of line quality through the monitoring of frame bits. Another limitation of the D4 format is the fact that obtaining a communications capability between devices on a T1 circuit required the use of a DS0 time slot. To alleviate these problems, as well as to provide the T1 user with additional capability, AT&T introduced an extended superframe format in early 1985 when it also introduced its D5 channel bank. Although this new framing format requires the installation of equipment that supports the frame format which slowed its introduction, within a few years a majority of T1 circuits in North America conform to this framing format. Eventually, the extended superframe format can be expected to completely replace D4 framing.

Denoted as Fe and ESF, the extended superframe format extends D4 framing to 24 consecutive frame bits—F1 through F24—as illustrated in Figure 6.5.

Unlike D4 framing in which the 12 framing bits form a specific repeating pattern, the ESF pattern can vary. ESF consists of three types of frame bits.

Derived data link

The ESF 'd' bits, which represent a derived data link, are used by the communications carrier to perform such functions as network

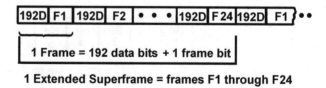

1 Frame = 192 data bits + 1 frame bit

1 Extended Superframe = frames F1 through F24

Figure 6.5 The extended superframe

monitoring to include error performance, alarm generation and reconfiguration to be passed over a T1 link. The 'd' bits appear in the odd frame positions, e.g., 1, 3, . . . , 21, 23. Since they are used by 12 of the 24 framing bits, the 'd' bits represent a 4 kbps data link.

As we will shortly note in this section, there are two ESF standards. One is actually a *de facto* standard defined in AT&T's Publication 54016. The second represents a *de jure* standard defined by the American National Standards Institute. Although both standards rely on the previously described ESF framing format, they differ with respect to the type of performance data recorded as well as in the manner by which they store and retrieve performance data.

The first standard was defined in AT&T's Publication 54016. In that publication the data link formed by the 12 'd' frame bits is coded into higher level data link control (HDLC) protocol format known as BX.25. Figure 6.6 illustrates the data link format carried by the 'd' bits.

The flag byte consists of the eight-bit sequence 01111110 and initiates and terminates each frame. The address field is used to identify a frame as either a command or a response. A command frame contains the address of the device to which the command is being transmitted, while a response frame contains the address of the device sending the frame.

The control field identifies the purpose of the frame and can indicate one of three frame types — supervisory, unnumbered or information. A supervisory frame is used for data link house-keeping information, such as acknowledgments. An unnumbered frame is used for major system commands, line initialization and shutdown information, while an information frame contains user data. The frame check sequence (FCS) is a 16-bit CRC check used to ensure the integrity of the data link.

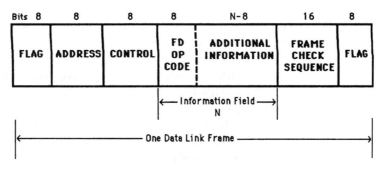

Figure 6.6 ESF data link format

One of the primary goals in the development of the BX.25 protocol was to provide a mechanism to extract performance information from ESF-compatible CSUs. Doing so allows circuit quality monitoring without taking the circuit out of service and is a major advantage of ESF over D4 framing. Standard maintenance messages that are defined in AT&T's publication 54016 include messages that can return performance data concerning the Out-of-Frame (OOF) conditions in which two out of four or five framing bits are in error, the number of errored seconds (ES), severely errored seconds (SES) and failed seconds (FS). An errored second is a second that contains one or more bit errors, while a severely errored second is considered to be a second with 320 or more bits in error. If 10 consecutive severely errored seconds occur, this condition is considered as a failed signal state. Then each signal in a failed state is considered to be a failed second. Table 6.1 summarizes the standard maintenance messages transmitted on the ESF data link based upon AT&T Publication 54016.

Error check link

Frame bits 2, 6, 10, 14, 18 and 22 are used for a CRC-6 code. The six-bit cyclic redundancy check sum is used by the receiving equipment to measure the circuit's bit error rate and represents 2 kbps of the 8 kbps framing rate. The CRC employs a mathematical algorithm which is used to check all 4632 bits in the ESF. Mathematically, the CRC check bit generation is performed by the use of a fixed polynomial whose composition is $X_6 + X + 1$, or 1000011. The data block is first multiplied by X_6 or 1 000 000 and then divided by the polynomial. The remainder is then transmitted in the six ESF CRC bit positions. At the receiver, a similar operation is performed on the received data block using the same polynomial and the locally generated check bit sequence is then compared to the received check bit sequence. The CRC-6 code yields an accuracy of 98.4%, and the occurrence of a mismatch between the locally generated check bit sequence and the received check bit sequence indicates that one or more bits in the extended superframe are in error.

To conform with AT&T's ESF CRC-6 coding and reporting requirements, the use of an ESF compatible channel service unit is required. This CSU not only generates the CRC-6 but must also be capable of detecting CRC errors and storing a CRC error count over a 24-hour period. ESF-compatible CSUs contain buffer storage which enables the device to store current line status information, including all error events and errored and failed seconds for the current 15-minute period and the previous 96

Table 6.1 ESF data link maintenance messages.

Send one-hour performance data
Upon receiving this command the ESF CSU will supply the following:
current status, elapsed time of current interval,
ES and FS in the current 15-minute interval,
number of valid intervals, count of ES and FS in 24-hour register,
ES and FS during the previous four 15-minute intervals.

Send 24-hour 'ES' performance data
Upon receiving this command, the ESF CSU will supply the following:
current status, elapsed time of current interval,
ESs and FSs in current 15-minute interval,
number of valid intervals, count of ESs and FSs in 24-hour register,
and ESs during prevous 96 15-minute intervals.

Send 24-hour 'FS' performance data
Upon receiving this command, the ESF CSU will supply the following:
current status, elapsed time of current interval,
ESs and FSs in the current 15-minute interval, number of valid intervals, count of
ESs and FSs in 24-hour register, and FSs during previous 96 15-minute intervals.

Reset performance monitoring counters
Upon receiving this command, the ESF CSU will reset all interval times and ES
and FS registers and supply the current status.

Send errored ESF data
Upon receiving this command, the ESF CSU will supply current data present in
ESF error event registers. Each count represents one error event (65 535
maximum).

Reset ESF register
Upon receiving this command, the ESF CSU will reset the ESF error event register
and supply the current status.

Maintenance loop-back (DLB)
Energizes upon receiving the proper code embedded in the 4 kbps data link. This
loop-back loops through the entire CSU.

15-minute periods that represent the prior 24-hour period. To
enable the carrier to retrieve this data, as well as to reset any or all
counters and activate or deactivate loop-back testing on the local
span line, the CSU must also have the ability to respond to network
commands. Thus, an AT&T Publication 54106 ESF-compatible
CSU must have the capability to send and receive data based upon
the BX.25 formation via the 4 kbps data link.

Framing pattern

Frame bits of the third type are used to generate the framing
pattern. Here, frame bits 4, 8, 12, 16, 20 and 24 are used to

Table 6.2 ESF framing pattern.

Frame	Bit composition	Frame	Bit composition
1	d	13	d
2	C1	14	C4
3	d	15	d
4	0	16	0
5	d	17	d
6	C2	18	C5
7	d	19	d
8	0	20	1
9	d	21	d
10	C3	22	C6
11	d	23	d
12	1	24	1

d = data link, Cx = CRC-6 bit x.

generate the Fe framing pattern whose composition is 001011. These six bits result in a 2 kbps framing pattern. Another difference between the ESF and D4 frame format is in the area of signaling. ESF has added two additional signaling bits, C and D, in frames 18 and 24. Thus, in ESF signaling data is accommodated by using bit robbing in frame 6 (A-bit), frame 12 (B-bit), frame 18 (C-bit) and frame 24 (D-bit). In Table 6.2 you will find a summary of the ESF framing pattern.

The ANSI T1E1 ESF standard

Although AT&T developed the original ESF framing method, its publication 54016 is not the only standard governing ESF framing. A second standard was developed by the American National Standards Institute (ANSI) T1E1 committee (T1.403-1989). Although both AT&T and ANSI standards are based upon the previously described ESF frame format, they differ from one another in the manner by which performance data is obtained and maintained.

Both ESF standards are communications carrier oriented as they provide a mechanism for the transfer of performance data from the subscriber to the network operator. Since it is important to enable your carrier to monitor the level of performance on your T1 connection, it is also important to use a CSU that is compatible or set to a compatible node of operation with the standard used by your network operator. Table 6.3 provides a summary of the similarities and differences between AT&T Publication 54016 and the ANSI T1.403-1989 standard.

Table 6.3   T1 performance standards.

Attribute	AT&T Publication 54016	ANSI T1.403-1989
8 kbps framing allocation	2 kbps framing 2 kbps CRC-6 4 kbps data link	2 kbps framing 2 kbps CRC-6 4 kbps data link

Monitored parameters	CRC-6 errors
	Out-of-Frame (OOF)
		CRC-6 errors
		Severely Errored Seconds
		  (same as OOF)
		Control Slip (optional)
		Framing Bit Error (optional)
		Bipolar Violations
		  (optional)

Derived parameters	Errored Second (ES) Severely Errored Second (SES) Failed Second (FS)	Range of CRC-6 errors Severely Errored Seconds Frame bit errors Bipolar violations Slip events
Historical performance duration	24 hours	3 seconds

8 kbps framing allocation

In examining the entries in Table 6.3 note that both standards use exactly the same allocation process for the 8 kbps frame sequence. As previously mentioned, their differences are in the manner by which performance data is obtained and maintained.

Monitored parameters

Under AT&T Publication 54016 only two parameters have to be monitored: CRC-6 errors where the calculated CRC-6 does not equal the received CRC-6, and Out-of-Frame (OOF) conditions. Under the T1E1 standard there are five error events, of which two are mandatory and three optional. Error events that must be detected are the same as under the AT&T standard; however, OOF was renamed as Severely Errored Framing under T1E1. Events whose detection is optional include Controlled Slips which represent the duplication or deletion of a complete frame, Framing Errors which represent the number of times a single framing bit was in error, and a Bipolar Violation (BPV), where a mark has the same polarity as the previous mark.

Derived parameters

As indicated in Table 6.3, AT&T Publication 54016 specifies three types of derived parameters based upon a second of time duration. Error Seconds (ES) represent a second having one or more CRC-6 errors and/or an OOF state. A Severely Errored Second (SES) represents a second having 320 or more CRC-6 errors and/or an OOF state. The third derived parameter, a Failed Second (FS), represents a second during which 10 consecutive SESs occur.

Under AT&T Publication 54016 the previously described derived parameters are organized into a series of performance registers that can be accessed via the data link from monitoring equipment installed in a carrier's central office. In comparison, the T1E1 standard does not define the need for equipment to extract parameters. Instead, under the T1E1 standard a performance report is transmitted every second.

Performance reports

Under the AT&T standard two derived parameters, Errored Seconds and Failed Seconds, are organized into 24-hour and one hour registers. Each register is further organized into 15-minute increments and all ESF errors, including CRC-6 and OOF states, are stored in a separate ESF Error Counter. The four registers representing hourly and 15-minute counts for Errored Seconds and Failed Seconds and the ESF Error Counter are extracted via the use of commands transmitted using the previously described BX.25 protocol.

In comparison to extracted performance reports under AT&T Publication 54016, the ANSI T1E1 standard specifies a performance report broadcast every second. That report consists of 112 bits which contain information about the current second and the three previous seconds. For each second information is reported concerning CRC-6 errors, Severely Errored Framing, frame bit errors, Bipolar Violations and slip events. Concerning CRC-6 errors, instead of the actual number of errors a range value is reported. Range values of CRC-6 errors are 1, 2–5, 6–10, 11–100, 101–319, and 320 or more.

Another difference between the two ESF standards concerns the manner by which performance data is transported. Instead of using a modified X.25 protocol the T1E1 standard specifies the use of the Q.921/Link Access Procedure D (LAPD) protocol.

Because the AT&T standard predates the ANSI T1E1 standard, initially CSUs supported only the former standard. However, most Regional Bell Operating Companies (RBOCs) and other carriers began to implement the ANSI standard, resulting in many manufacturers building equipment to support both standards. In fact, due to the brevity of the T1E1 performance report which requires 112 bits per second from the 4 kbps data link portion of the ESF framing signal, it is possible to broadcast T1E1 data as well as extract on demand AT&T register stored data.

Although AT&T made a substantial investment in Publication 54016 ESF-compatible equipment several years ago, it also made a commitment to support the ANSI standard. This means you can reasonably expect to be able to obtain ANSI T1E1 support regardless of the communications carrier you plan to use.

T1 alarms and error conditions

There are several alarms and error conditions that are monitored and reported under the T1 D4 and ESF formats. Principal T1 alarms include a red alarm which is produced by a receiver to indicate that it has lost frame alignment, and a yellow alarm which is returned to a transmitting terminal to report a loss of frame alignment at the receiving terminal. Normally, a T1 terminal will use the receiver's red alarm to request that a yellow alarm be transmitted.

To illustrate the operation of T1 alarms, consider the configuration illustrated in Figure 6.7 that shows two channel banks connected together via the use of a four-wire T1 circuit. If the second channel bank (CB2) loses synchronization with the first channel bank (CB1) or completely loses the signal, it then transmits a red alarm to CB1. The transmission of the red alarm is accomplished by CB2 forcing bit position 2 in each PCM word to

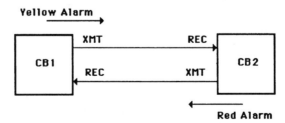

Figure 6.7 T1 alarm generation example. A red alarm is produced by a receiver to indicate it has lost frame alignment, while a yellow alarm is returned to a transmitting terminal to report a loss of frame alignment at the receiving terminal

zero and the frame signaling bit in frame 12 to a binary 1 under the D4 format. If ESF framing is used, CB2 sends a repeated pattern of eight 0s and eight 1s on the data link.

When CB1 recognizes the red alarm, it transmits a yellow alarm. This alarm, in effect, says 'I'm sending data but the other end is not receiving the transmitted information and the problem is elsewhere.' Under D4 framing the yellow alarm is generated at the receiver by setting bit 2 to 0 for 255 consecutive channels and the frame alignment signal (Fs) to 1 in frame 12. Under ESF a pattern of eight 0s and eight 1s repeated 16 times is used to indicate a yellow alarm.

A red alarm is generated when either the network or a DTE senses an error in the framing bits for either two out of four or five framing bits and this condition persists for more than 2.5 seconds. Table 6.4 summarizes the method of transmitting alarms on D4 and ESF T1 circuits.

In addition to red and yellow alarms, a third type of alarm that warrants attention is a blue alarm. This alarm, which is also known as an alarm indicating signal (AIS), is generated by a higher order system (HOS), such as a T1C system operating at 3.152 Mbps.

The T1C system was introduced by AT&T in 1975 and uses an asynchronous multiplexer to generate a DS1C signal. That signal combines two DS1 signals representing 48 DS0s. Although each DS1 signal operates at 1.544 Mbps, a DS1C signal uses pulse stuffing to compensate for variations in each DS1 input signal, resulting in the composite DS1C signal operating at 3.152 Mbps. Later in this chapter when we examine the T3 carrier, we will also examine the manner by which bit stuffing is employed to compensate for variations in two or more asynchronous signals that require multiplexing.

Figure 6.8 illustrates the generation of blue alarms when a failure between higher order systems occurs. This alarm is generated by each HOS system to channel banks or multiplexers and, in effect, tells each lower ordered system (DTE) that the problem is between higher ordered systems. Thus, this alarm is

Table 6.4 T1 alarm formats.

Mode	Format
Transmitted red alarm	
T1 D4	Bit 2 = 0 in all data channels and Fs = 1 in frame 12
T1 ESF	Repeated pattern of eight 0s, eight 1s, on data link
Yellow alarm generated at receiver	
T1 D4	Bit 2 = 0 for 255 consecutive channels and Fs = 1 in frame 12
T1 ESF	16 patterns of eight 0s, eight 1s, on data link

primarily used by communications carriers and avoids the dispatching of maintenance personnel to lower order system facilities. Normally, a blue alarm is generated after 150 ms of loss of an incoming signal. This alarm is produced by transmitting a continuous 1s pattern across all 24 channels.

Note that the previously discussed T1 alarms are not implemented on all T1 circuits. As an example, a point-to-point T1 circuit that is directly routed between two locations and is not switched nor multiplexed by a communications carrier will not generate the previously discussed alarms. However, if the end-user installs equipment that is capable of recognizing alarm conditions and generating those alarms, the T1 line will then pass those alarms generated by their equipment.

In addition to the previously described alarm conditions, there are several error conditions that can be detected by appropriate equipment connected to a T1 circuit. These error conditions include loss of carrier, bipolar violations, and Fs and Ft bit errors.

A loss of carrier condition is defined when receive data is zero for 31 consecutive bits. A bipolar violation is a failure to meet the AMI T1 line code in which marks (1s) are transmitted alternately as positive or negative pulses, while zeros are transmitted as zero volts. An Fs bit error indicates that a signaling framing bit is in error, while an Ft bit error indicates that a terminal framing bit is in error.

CEPT PCM-30 format

CEPT PCM-30, which is commonly referred to as an E1-carrier, is a PCM format used for time division multiplexing of 30 voice or data circuits onto a single twisted-pair cable using digital repeaters. Each voice circuit is sampled at 8 kHz using an eight-bit A-law

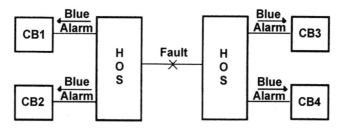

Figure 6.8 Blue alarm generation: HOS = higher order system, CB = channel bank

companding analog-to-digital converter and multiplexed with 29 other sampled channels plus one alignment and one signaling channel, resulting in 32 multiplexed channels.

The standard CEPT frame is 32 channels×8 bits/channel or 256 bits. With 8000 samples per second, the CEPT data rate becomes 8000×256, or 2.048 Mbps. Note that under this format there are no framing bits added to a frame as is the case under the North American T1 format. This is because the framing bits are carried within specific time slots.

Frame composition

Each CEPT PCM-30 frame consists of 32 time slots, comprising 30 voice, one alignment and one signaling, with each time slot represented by eight bits. Since each PCM channel is sampled 8000 times per second, the standard CEPT-30 data rate is 32×8×8000, or 2.048 Mbps.

Alignment signal

An alignment signal (0011011) is transmitted in bit positions 2 to 8 of time slot 0 in alternating frames. This signal is used to enable each channel to be distinguished at the receiver. Bit position 1 in time slot 0 carries the international bit, while frames not containing the frame alignment signal are used to carry national and international signaling and alarm indication for loss of frame alignment. Figure 6.9 illustrates the composition of the CEPT-30 frame and multiframe, where the multiframe consists of 16 frames, numbered from frame 0 to frame 15.

To avoid imitation of the frame alignment signal, alternating frames fix bit 2 to a 1 in time slot 0 which is the reason why a 1 is entered into that bit position for odd time slot 0 frames.

Signaling data

Time slot 16 in each CEPT-30 frame is used to transmit such signaling data as on-hook and off-hook conditions, dialing digits and call progress. This is indicated by the characters ABCD for frames 1 to 15 in time slot 16 illustrated in Figure 6.9. Since a common channel is dedicated for the signaling data of all-voice circuits, this method of signaling is referred to as common channel signaling and enables each time slot to operate at a 64 kbps. In comparison, T1 uses bit robbing to pass signaling information, which reduces the effective data rate of time slots to 56 kbps. To enable each frame in a multiframe to be distinguished at the receiver for the recovery of ABCD signaling, CEPT-30 uses a multiframe alignment signal. This signal, denoted by the symbol MAS in Figure 6.9, is transmitted in bit positions 1 through 4 of time slot 0 of frame 0. The use of each of the 32 CEPT PCM-30 time slots and their numbering is summarized in Table 6.5.

CEPT alarms and error conditions

The principal alarms defined by the CEPT PCM-30 format include a red alarm which is produced by a receiver to indicate that it has lost frame alignment and a yellow alarm which is returned to the transmitting terminal to report a loss of frame alignment at the receiving terminal. Both of these alarms function the same, as previously described in the section covering the T1 carrier.

Both red and yellow alarms are generated through the use of the alarm indication signal bit (bit 3) in time slot 0 (TS0) of odd frames. A red alarm is generated by setting bit 3 = 1 in TS0 of non-frame alignment frames. A receiver then indicates the reception of a red alarm by generating a yellow alarm by setting bit 3 = 1 in TS0 of non-frame alignment frames.

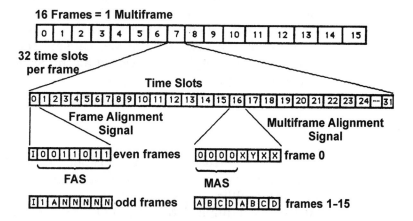

Figure 6.9 CEPT PCM-30 frame and multiframe composition: I = international bit, N = national bit, A = alarm indication signal, FAS = frame alignment signal, ABCD = ABCD signaling bits, X = extra bit for signaling, Y = loss of multiframe alignment, MAS = multiframe alignment signal

Table 6.5   CEPT PCM-30 time slot and channel numbering.

Time slot	Channel	Data use	Time slot	Channel	Data use
0	FAS	No	16	MAS	No
1	1	Yes	17	16	Yes
2	2	Yes	18	17	Yes
3	3	Yes	19	18	Yes
4	4	Yes	20	19	Yes
5	5	Yes	21	20	Yes
6	6	Yes	22	21	Yes
7	7	Yes	23	22	Yes
8	8	Yes	24	23	Yes
9	9	Yes	25	24	Yes
10	10	Yes	26	25	Yes
11	11	Yes	27	26	Yes
12	12	Yes	28	27	Yes
13	13	Yes	29	28	Yes
14	14	Yes	30	29	Yes
15	15	Yes	31	30	Yes

FAS = Frame alignment signal, MAS = multiframe alignment signal.

Two additional alarms generated by CEPT PCM-30 include a multiframe red alarm and a multiframe yellow alarm. The multiframe red alarm is produced by a receiver to indicate that it has lost the multiframe alignment, while the multiframe yellow alarm is returned to the transmitting terminal to report a loss of frame alignment at the receiving terminal. A receiver loses multiframe alignment either if two consecutive errors occur in the multiframe alignment signal or if time slot 16 contains all zeros for at least one multiframe, causing the red alarm to go high, which is coupled to a yellow alarm generator (bit $6 = 1$ in time slot 16, frame 0).

CEPT CRC option

For enhanced error monitoring capability, CEPT PCM-30 includes a CRC-4 option. Under this option, a group of eight frames known as a submultiframe is treated as a long binary number. This number is multiplied by X^4 (10000) and divided by $X^4 + X + 1$ (10011). The four-bit remainder is transmitted in bit position 1 (I bit in Figure 6.9) in time slot 0 in even frames which contain the frame alignment signal. After the receiver computes its own CRC-4 check, it uses bit position 1 in time slot 0 of frames 13 and 15 for

Table 6.6 CEPT PCM-30 CRC error performance reporting.

Bit 1 Frame 13	Bit 1 Frame 15	
1	1	CRC for SMF I, II error free
1	0	SMF II in error, SMF I error free
0	1	SMF II error free, SMF I in error
0	0	Both SMF I and II in error

CRC error performance reporting. Table 6.6 summarizes how these bits are used for CRC error performance reporting purposes.

The principal CEPT PCM-30 error conditions include the occurrence of bipolar violations, frame alignment errors and multiframe alignment errors. A bipolar violation is a failure to meet the AMI CEPT PCM-30 line code, where marks alternate as positive and negative pulses and spaces are represented by a zero voltage. A frame alignment error is a failure to synchronize on the frame alignment signal (0011011) contained in time slot 0 of alternating frames, while a multiframe alignment error is a failure to synchronize on the multiframe pattern (0000) contained in bits 1–4 of time slot 16 of frame 0.

6.2 T1- AND E1-CARRIER SIGNAL CHARACTERISTICS

As previously discussed in Chapter 1, there are several advantages to transmitting T- and E-carrier signals in a bipolar alternate mark inversion (AMI) format, including the absence of a dc component to the signal and the ability to obtain clock recovery from the signal in an all-1s condition. Unfortunately, the disadvantage associated with this signaling method is that a sequence of spaces is encoded as a period of zero voltage or no signal, and repeaters on a span line cannot recover clocking without a signal occurring every so often.

For repeaters to properly recover clocking, a certain number of binary 1s must be contained within a transmitted signal. Since North American and European approaches to this problem differ, we will examine the encoding methods used to ensure that an appropriate signal contains a minimum number of binary 1s for each location. This examination will be focused on the T1- and E1-carriers in this section. Since we will examine the signal characteristics of the T3-carrier as a separate entity in the next section in this chapter, we will defer a discussion of the manner by which a minimum 1s density is maintained on a T3-carrier until that section.

North America

In North America, AT&T publication 62411 sets the 1s density requirement to be n 1s in each window of $8 \times (n + 1)$ bits, where n varies from 1 to 23. This means that a T1-carrier cannot have more than 15 consecutive 0s ($n = 1$) and there must be approximately three 1s in every 24 consecutive bits ($n = 2$ to 23). Several methods are currently used to provide this minimum 1s density, including binary 7 zero code suppression, binary 8 zero substitution, and zero byte time slot interchange.

Binary 7 zero code suppression

Under the binary 7 zero code suppression method a binary 1 is substituted in bit position 7 of each time slot if all eight positions are zeros. An example of this method of ensuring a minimum 1s density is illustrated in Figure 6.10A. Although it might appear wiser to select the least significant bit for inversion, this cannot be done since the setting of a frame bit to 0, when bit positions 2 through 8 in the previous time slot were set to 0, would result in a string of 16 consecutive 0s if the bits in the time slot following the frame bit were 0 and bit position 8 was used for substitution. Figure 6.10B illustrates this worst-case scenario which explains why bit position 7 in each time slot is used for bit value inversion to ensure a minimum 1s density.

Effect upon bit robbing

If a data channel contains all 0s, the data can be corrupted due to B7 zero suppression. Due to this, a data channel normally is restricted to seven usable data bits, with one bit in the data channel set to a 1. This prevents the data channel from being corrupted, but also limits its data rates to 56 kbps.

When one bit is set to a 1 on a DS0 channel, the channel is known as a non-clear channel. The 56 kbps on a non-clear channel is also known as a DS-A channel.

A T1 clear channel is one in which all 64 kbps in each DS0 are usable. On private microwave systems, as well as when transmission occurs over fiber, B7 zero code suppression is normally not required, permitting clear channel capability.

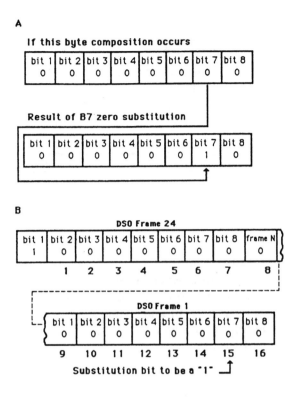

Figure 6.10 B7 zero code suppression: A B7 zero code suppression example; B worst-case scenario

Binary 8 zero substitution

The binary 8 zero substitution (B8ZS) technique was developed by Bell Laboratories and is now sanctioned by the ITU for use in North America. This method of ensuring a minimum 1s density was placed into operation during the mid-1980s and offers a significant improvement over binary 7 zero code suppression, as it both maintains a minimum 1s density and also provides a clear channel capability, permitting each DS0 channel to carry data at 64 kbps. Under B8ZS coding, each eight consecutive 0s in a byte are removed and replaced by a B8ZS code. If the pulse preceding an all-0s byte is positive, the inserted code is $000 + -0 - +$. If the pulse preceding an all-0s byte is negative, the inserted code is $000 - +0 + -$. Figure 6.11 illustrates the use of B8ZS coding in which an all-0s byte is replaced by one of two binary codes, with the actual code used based upon whether the pulse preceding the all-0s byte was positive or negative.

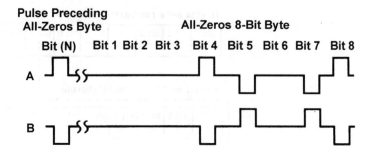

Figure 6.11 B8ZS coding

Both examples result in bipolar violations occurring in the fourth and seventh bit positions. Both carrier and customer equipment must recognize these codes as legitimate signals, and not as bipolar violations or errors, for B8ZS to work to enable a receiver to recognize the code and restore the original eight 0s.

Zero byte time slot interchange

Zero byte time slot interchange (ZBTSI) is a method of encoding information into a PCM word to prevent an excessive number of zeros from occurring on the transmission line. Unlike the two previously discussed zero suppression methods that can be used in any framing method, ZBTSI is available only with an extended superframe format.

When a PCM word containing all zeros is found, a CSU with a ZBTSI encoder will replace the zero byte with addressing information, while overhead information is transmitted by using 2 kbps of the 4 kbps ESF data link. The encoder operates on 96 PCM words at a time, representing four ESF frames. The data link bits from frames 1, 5, 9, 13, 17 and 21 are used to provide encoding flags for the 96 words following each flag bit and are referred to as Z bits. Figure 6.12 illustrates the relationship of the ZBTSI flag bits with respect to the ESF frames.

ZBTSI represents a proprietary 1s density technique developed by Verilink, a manufacturer of CSUs. The ZBTSI method is supported by several Bell Operating Companies and many modern CSUs support both B8ZS and ZBTSI.

ESF Frame Number	ESF Bit Number	Framing Bits Assignments			
				Data Link	
		Fe	DL	ZBTSI	CRC
1	0	-	-	z	-
2	193	-	-	-	CB1
3	386	-	d	-	-
4	579	0	-	-	-
5	772	-	-	z	-
6	965	-	-	-	CB2
7	1158	-	d	-	-
8	1351	0	-	-	-
9	1544	-	-	z	-
10	1737	-	-	-	CB3
11	1930	-	d	-	-
12	2123	1	-	-	-
13	2316	-	-	z	-
14	2509	-	-	-	CB4
15	2702	-	d	-	-
16	2895	0	-	-	-
17	3088	-	-	z	-
18	3281	-	-	-	CB5
19	3474	-	d	-	-
20	3667	1	-	-	-
21	3860	-	-	z	-
22	4053	-	-	-	CB6
23	4246	-	d	-	-
24	4439	1	-	-	-

Figure 6.12 ESF framing bit assignments with ZBTSI: Fe = framing pattern sequence, DL = 4 kbps data link channel, ZBTSI = ZBTSI encoding flag bits (Z bits), CRC = cyclic redundancy check field

Europe

In Europe the high density bipolar 3-zero maximum (HDB3) coding is used by CEPT PCM-30 to obtain a minimum density for clock recovery from received data.

HDB3 coding

Under HDB3, the data stream to be transmitted is monitored for any group of four consecutive 0s. A four-zero group is then replaced with an HDB3 code. Two different HDB3 codes are used to ensure that the bipolar violation pulses from adjacent four-zero groups are of opposite polarity as indicated in Figure 6.13. The selection of the HDB3 code is based upon whether there was an odd or even number of 1s since the last bipolar violation (BV) occurred. If an odd number

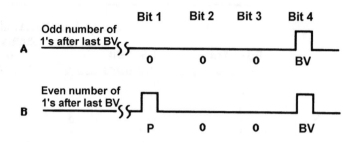

Figure 6.13   HDB3 coding: P = polarity bit, BV = bipolar violation

of 1s occurred since the previous bipolar violation, the coding method in Figure 6.13A is used to replace a sequence of four zeros. If an even number of 1s occurred since the previous bipolar violation, the coding method in Figure 6.13B is used to replace a sequence of four zeros.

6.3 THE T3-CARRIER

Although many organizations think of the T3-carrier as a modern digital transport system, it represents an AT&T-developed standard that has been in use for approximately 30 years. Until the mid-1990s the relatively large transmission capacity of the T3-carrier as well as its cost placed it beyond the practical use of most organizations. The growth in the use of the Internet resulted in many popular Web sites experiencing millions of user accesses per day which created a demand for transmission bandwidth that could only be satisfied through the use of a T3 transmission facility. In addition, Internet Service Providers as well as commercial organizations and government agencies found the use of T3-carriers ideal for consolidating transmission, enhancing its popularity. The growth in the construction of fiber optic long-distance transmission facilities resulted in a significant amount of bandwidth becoming available for use. This resulted in competitive pressure among long-distance communications carriers to fill up available bandwidth. To do so they began to offer significant discounts and price reductions which further enhanced the popularity of T3 circuits.

T3 circuit types

There are two types of T3 circuits available for use. The first type of T3 circuit was structured to transport multiple levels of multiplexed digital voice signals using a two-stage multiplexing scheme.

This type of T3 circuit is referred to as a channelized or subrated T3. As we will shortly note, the channelized or subrated T3 circuit is designed to transport the equivalent of 28 T1 circuits.

A second type of T3 is a non-channelized T3 circuit. This type of T3 circuit does not support multiple levels of multiplexed digitized voice signals. However, it uses every 85th bit as a framing bit, resulting in a 44.210 Mbps transmission capacity on the 44.736 Mbps signal.

Evolution

Although the 1.544 Mbps T1 circuit significantly reduced cable congestion in urban areas, its operating rate was not sufficient for certain high volume locations. In 1972 AT&T implemented testing of what is now the second level of the digital transmission hierarchy in North America which is referred to as a DS2 signal.

The DS2 signal

The DS2 signal is formed through the use of an M12 multiplexer which combines four asynchronous DS1 signals (96 DSOs) for transmission over a synchronous link. Since the DS1 signals are asynchronous, the multiplexer required a method to compensate for the variations in the clock rates among the DS1 signals. The method of clock compensation selected by AT&T is referred to as pulse stuffing which is effected by the addition of another layer of framing beyond DS1 framing to the DS2 signal. This additional layer of framing provides for the correct alignment of bytes in a manner similar to DS1 framing; however, it also enables variations in the clocking of DS1 signals through the use of stuffing bits. Such stuffing bits are added as needed to ensure each DS1 signal has an identical bit rate prior to the M12 multiplexer bit interleaving the signals. A relatively 'fast' signal will receive less stuffing, while a relatively 'slow' signal will receive more stuffing. At the far end of the transmission path the stuffing bits are removed.

Once the DS2 hierarchy was in operation its 6.312 Mbps proved insufficient for many applications. A further development resulted in the output of seven M12 multiplexers being re-multiplexed into one composite high speed circuit referred to as a T3-carrier. This second-stage level of multiplexing is performed by an M23 multiplexer, whose acronym refers to layer 2 in the digital hierarchy input to provide a layer 3 signal.

Figure 6.14 illustrates the construction of a DS3 signal as a result of a two-stage multiplexing process. Since an understanding of T3 framing requires an understanding of T2 or DS2 framing, let's first focus our attention on that topic.

DS2 framing

A DS2 frame is 1272 bits long, representing the multiplexing of four DS1 signals. Within the DS2 frame are four subframes, each 318

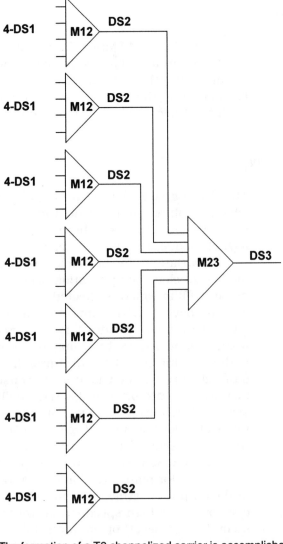

Figure 6.14 The formation of a T3 channelized carrier is accomplished via a two-stage multiplexing process

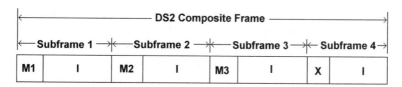

I = 293 information bits
M = Multiframing bits
X = User or application set

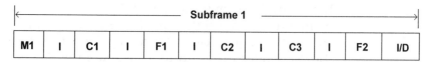

M = Multiframe bits
F = Framing bits
C = Control bits
I = 48 information bits
I/D = Inserted/Data bit

Figure 6.15 DS2 framing results in the creation of four subframes that are multiplexed together by time

bits in length. This is illustrated in the top portion of Figure 6.15. Each subframe is further divided into six blocks of 53 bits as illustrated in the lower portion of Figure 6.15.

If we examine Figure 6.15 from the bottom up we can obtain a better appreciation for the manner by which framing is accomplished. At the first stage of framing an alternating 01 bit pattern is formed by the F bits which occur every 159 bits in the composite DS2 frame. The second framing stage results from the use of Multiplexing (M) bits which reside in the first bit position of each subframe. The sequence of M bits form the pattern 011X, with the 01 transition used to identify the beginning of the DS2 master frame. The X bit position can be set to either 0 or 1 and reflects a user or application value.

Pulse stuffing

As previously noted, stuffing bits are used to synchronize four DS1 signals to compensate for clocking differences between each signal. The fact that stuffing bits are employed resulted in the requirement to use control bits. Thus, the C bits in the subframe are used to identify the contents of the stuffing bits. That is, if a majority of the C bits (two out of three) are set to 1 the stuffing bit is holding an inserted (I) bit, otherwise the stuffing bit contains data. Since the stuffing bit may or may not contain data, each subframe can contain either 311 or 312 data bits.

Maintaining ones density

Earlier in this chapter we noted that one common method used to maintain a 1s density on a T1 line is B8Zs. A DS2 signal uses a similar technique, referred to as Bipolar with Six Zero Substitution (B6ZS). Under B6ZS up to a maximum of five consecutive zeros are permitted prior to a six-bit byte to include intentional bipolar violations being substituted for a sequence of six consecutive zeros.

T3 framing

As previously indicated, the formation of a T3 channelized carrier is accomplished via a two-stage multiplexing process. In the first stage four DS1 signals are combined into a DS2 signal. During the second stage seven DS2 signals are multiplexed to form a DS3 signal. Although Figure 6.14 illustrated two separate multiplexers as being used to form the two-stage multiplexing process, that process can also occur as the result of a single multiplexer. That multiplexer is referred to as an M13 device as it takes 28 DS1 signals, and although it internally first forms seven DS2 signals, it multiplexes those signals with the device to form one DS3 signal. Regardless of the number of separate multiplexers used to form a DS3, the resulting signal has a specific framing format. In actuality, although there is only one framing format, the use of certain bits falls into one of two camps, which result in two standards. The first standard for T3 framing assumes that bit stuffing occurs at each stage in the multiplexing process and results in the M13 format. Since bit stuffing from T2 to T3 is unnecessary when one device performs T1 to T3 multiplexing, there is no need to use certain bits for bit stuffing. Recognition of this as well as the need for an enhanced maintenance capability resulted in the development of a new T3 framing format. That format, known as the C-bit parity format, was proposed by AT&T and ratified as ANSI standard T1.107A in 1990.

The M13 framing format

A DS3 signal is divided into frames 4760 bits in length that are referred to as M-frames. Since each M-frame represents seven multiplexed DS2 signals, it is no surprise that each M frame also consists of seven M-subframes. However, these subframes do not represent individual DS2 signals, as the DS3 signal is formed by bit-interleaving multiplexing of the seven DS2 signals. Since the M-frame is 4760 bits in length, each M-subframe is 680 bits in length.

Figure 6.16 illustrates the T3 M-frame and an M-subframe. Note that each M-subframe is further subdivided into eight blocks, each consisting of an overhead bit followed by 84 data bits.

In examining the M-Frame shown in Figure 6.16 note that the X bits were originally used to establish a signaling channel and prefix user information on the first two subframes used to form the M-frame. Today the X bits are used to indicate loss of signal and loss of frame. The P bits, which prefix the information bits in subframes 3 and 4, contain the Modulo 2 sum of data bits in the previous frame. The X bits and P bits are controlled by transmitting equipment to provide basic alarm and error indication information across a T3 line. The M bits that prefix the information bits in subframes 5 through 7 identify the relative subframe position. To

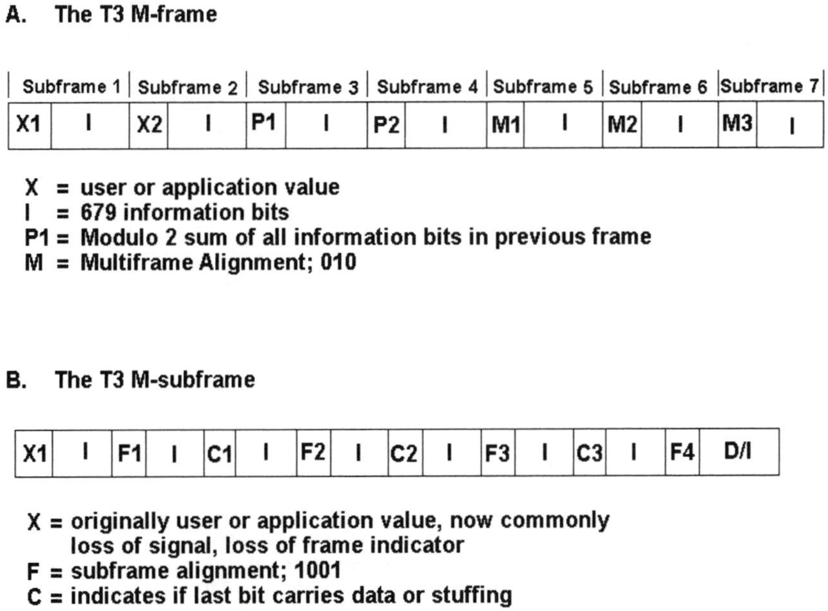

A.   The T3 M-frame

Subframe 1		Subframe 2		Subframe 3		Subframe 4		Subframe 5		Subframe 6		Subframe 7	
X1	I	X2	I	P1	I	P2	I	M1	I	M2	I	M3	I

X = user or application value
I   = 679 information bits
P1 = Modulo 2 sum of all information bits in previous frame
M = Multiframe Alignment; 010

B.   The T3 M-subframe

X1	I	F1	I	C1	I	F2	I	C2	I	F3	I	C3	I	F4	D/I

X = originally user or application value, now commonly
       loss of signal, loss of frame indicator
F = subframe alignment; 1001
C = indicates if last bit carries data or stuffing

Figure 6.16   The T3 M-frame and M-subframe

do so they transport the bit sequence 010. Because the M bits provide a frame alignment capability they are also cumulatively referred to as the MultiFrame Alignment signal.

The lower portion of Figure 6.16 illustrates the general composition of an M-subframe. The four F bits form a subframe alignment signal which has the repeating pattern '1001'. Similarly to the DS2 frame format, the three C bits indicate whether the last bit position in the subframe carries payload data or stuffing.

To facilitate a visual observation of the M13 framing process Figure 6.17 illustrates the M13 framing bit pattern. Note that the F bits are fixed in value to either a 1 (F1) or 0 (F0) value. Similarly, the M bits are also fixed in value to either a 1 (M1) or 0 (M0) value. Note that the control bits (C1, C2, C3) define the use of the last bit position in each subframe. That is, when two or more bits are set the last bit in the subframe is a stuffed bit, otherwise the last bit transports data.

The C-bit parity format

As previously noted in our examination of the M13 framing format, each DS3 signal is developed by combining seven subframes, each containing eight overhead bits. This results in each DS3 M-frame containing 58 overhead bits, of which 21 were used to indicate bit stuffing. The ability to use the 21 C bits for other purposes resulted in the development of the C-bit parity format. This framing format results in the use of the C-bits for performance monitoring and maintenance instead of as a mechanism to define whether or not the last bit in each subframe is carrying data or is a stuffing bit.

Subframe	Block #1							Block #8
1	X1	F1	C1	F0	C2	F0	C3	F1
2	X2	F1	C1	F0	C2	F0	C3	F1
3	P1	F1	C1	F0	C2	F0	C3	F1
4	P2	F1	C1	F0	C2	F0	C3	F1
5	M0	F1	C1	F0	C2	F0	C3	F1
6	M1	F1	C1	F0	C2	F0	C3	F1
7	M0	F1	C1	F0	C2	F0	C3	F1
	Block #49							Block #56

F = Subframe Alignment; 1001
C = Pulse Stuffing Indicator Codes
M = Multiframe Alignment; 010
X = Loss of frame, loss of signal indicators

Figure 6.17 The M-framing bit pattern

Block Number

Subframe

	1	2	3	4	5	6	7	8
1	X	F1	AIC	F0	N	F0	FEAC	F1
2	X	F1	DL_1	F0	DL	F0	DL	F1
3	P	F1	CP	F0	CP	F0	CP	F1
4	P	F1	FTBE	F0	FEBE	F0	FEBE	F1
5	M0	F1	DL_2	F0	DL	F0	DL	F1
6	M1	F1	DL	F0	DL	F0	DL	F1
7	M0	F1	DL	F0	DL	F0	DL	F1

AIC = Application Identification Channel =1

N = Network Application Bit (Reserved)

FEAC = Far End Alarm and Control Channel

DL = Data Link

CP = C-Bit Parity

FEBE = Far End Block Error

X = Degraded Second

Figure 6.18 The C-bit parity framing format

Figure 6.18 illustrates the DS3 C-bit parity framing format. Note that only the bits in blocks 3, 5 and 7 are changed from their use from the DS3 M13 frame format previously illustrated in Figure 6.17.

The Application Identification Channel indicates whether or not C-bit parity is being used. If so, framing bit 3 in subframe 1 is set to 1. The Far End Alarm and Control Channel (FEAC) bit is used for transmitting performance data as well as loopback commands on an end-to-end basis. Thus, the FEAC provides the ability to automatically place a T3 circuit into a loopback mode of operation.

The three CP bits represent path parity bits that are calculated by the transmitter. The receiver uses these C-bit parity bits to perform a parity check for transmission errors, enabling line quality statistics to be computed without having to take the line out of service. Information about received parity errors is returned to the transmitter via the use of three Far End Block Error (FEBE) bits.

One additional change resulting from the C-bit parity framing format concerns the use of the first framing bit in the first two subframes. Those X bits, which were used to indicate loss of frame or loss of signal under the M13 format, are used to indicate a degraded second under the C-bit parity framing format. When used in this manner, a degraded second simply denotes a second in which a Loss of Frame or Loss of Signal occurred.

Ones density

Similar to other types of T-carrier transmission systems, the DS3 signal needs to provide a minimum 1s density to enable repeaters to obtain clocking from pulses on the line. To accomplish this, Binary 3 Zero Substitution (B3ZS) coding is employed. Another difference between T3 and T1 and T2 is in their transmission media. While T1 and T2 are transported over twisted-pair cable, T3 is commonly transported over 75 ohm coaxial cable to the carrier's equipment. From the network interface the T3 is commonly transmitted over fiber optic cable or via microwave radio.

Access and network utilization

Unlike lower speed T-carriers, the use of a T3 transmission system can require a considerable degree of planning due to the special interfaces needed to support relatively high transmission rates. Although access to a T3 circuit is similar to the manner by which a T1 is accessed, the DSU/CSU used with a T3 needs to support a high speed interface.

Data rate interface

For data rates up to 6 Mbps you can use V.35, RS449 and EIA 530 interfaces. However, this means that unless you have multiple data sources and the DSU/CSU performs multiplexing, you will not be able to utilize the full transmission capacity of the T3 system. As an alternative or supplement to the previously mentioned data interfaces, vendors offer one or more High Speed Serial Interface (HSSI) connections. The HSSI supports data rates from 3 to 52 Mbps and can permit a more effective use of a T3 transmission system. In fact, many routers are now manufactured to support an optional HSSI data interface.

The top part of Figure 6.19 illustrates the use of a multiport T3 DSU/CSU to provide access to a T3 transmission system, while the lower portion of that figure indicates the use of a T3 multiplexer. Similar to a T1 DSU/CSU, the T3 CSU provides the electrical interface, signal formatting, performance monitoring, alarm signal generation and loopback operations, while the DSU provides the data interface. The network interface side of the CSU must support an operating rate of 44.736 Mbps. Due to the popularity of M13 and C-bit parity framing formats, most CSUs have the ability to be set to support either framing format. Table 6.7 provides a list of seven common T3 CSU network interface parameters.

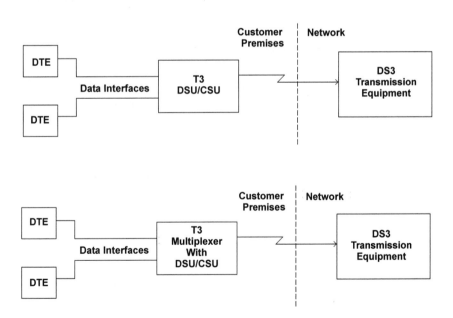

where Interfaces = V.35, RS449, EIA 530 for data rates up to 6 Mbps

HSSI for data rates from 3 to 52 Mbps

Figure 6.19 Common T3 access methods

Table 6.7 CSU T3 network interface parameters.

Line operating rate	DS3 44.736 Mbps
Framing format	M13 or C-bit parity
Input signal	+62 dBm to −11.7 dBm
Output signal	DS3
Impedance	75 ohm coax
Connector	BNC socket
Timing	Looped or internal clock

Network access

Although most ISPs use full T3 circuits on a point-to-point basis to connect to an Internet homing location, other organizations typically use T3 circuits either as access to a fractional T3 transmission facility or as a homing point for individual T1 circuits. When used as an access mechanism to a fractional T3 transmission service, a full T3 circuit is installed between the customer premises and the central office point-of-presence. The T3 access

Accessing fractional T3 service

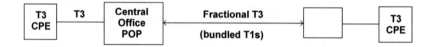

Common T1 access as a homing point or T3 fan-out

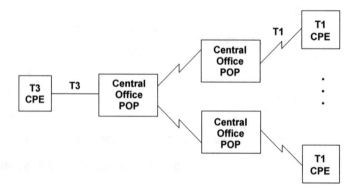

Figure 6.20 Common T3 access methods

line is configured as a combination of T1s in the same manner that a T1 represents a series of DS0s used to provide access to a fractional T1 service. This type of T3 is referred to as a subrated T3 and is illustrated in the top part of Figure 6.20.

When a T3 circuit is used as a homing point for multiple T1s routed to the same or different geographical areas, the T3 access line is also divided into combinations of T1s. Thus, the T3 access line is also a subrated transmission facility. Since the use of a single T3 access line to provide access to multiple T1 lines forms a fan-out network configuration, this network configuration is also commonly referred to as a fan-out.

When a T3 is used to provide access to fractional T3 service or as a homing point for multiple T1s, it is configured as a channelized or subrated signal. This means that the C-bit framing format cannot be maintained between customer premises; however, it can be used by the network operator. Thus, although you may not be able to obtain end-to-end performance measurements, the network operator can act upon your behalf and note the condition of this important transmission facility.

REVIEW QUESTIONS

1 How does loop timing provide synchronization on a point-to-point line connection?

2 What was the original purpose of the D1 framing bit?

3 Describe the composition of a DS1 frame, including the placement of each DS0 PCM word and the framing bit.

4 What is the actual data rate available for use on a DS1 signal?

5 How many bits are included in a superframe?

6 Discuss a key limitation associated with the use of D4 framing.

7 Discuss two advantages of the use of the extended superframe format over D4 framing.

8 How many bits are in an extended superframe?

9 Discuss the use of each of the three types of frame bits in the extended superframe.

10 What does a mismatch between the locally generated check bit sequence and the received check bit sequence under the extended superframe format indicate?

11 What are the key differences between the AT&T implementation of the extended superframe and the ANSI T1E1 standard?

12 Discuss the relationship between a yellow and a red alarm.

13 Where does a blue alarm originate and what does it indicate?

14 How many bits are in a CEPT frame? Where are the framing bits in a CEPT frame?

15 How many bits are contained in the CEPT multiframe?

16 Why can a CEPT DS0 carry data at 64 kbps, while a T1 DS0 is normally limited to a 56 kbps data rate?

17 Why was the seventh bit in a PCM word selected for inversion under binary 7 zero code suppression instead of the least significant bit?

18 What is the common advantage of binary 8 zero substitution and zero byte time slot interchange zero suppression methods over binary 7 zero code suppression?

19 What is the difference between a channelized and a non-channelized T3 circuit?

20 What method is used to compensate for the variations in clock rates among DS1 signals by an M12 or an M13 multiplexer?

21 Describe the two-stage multiplexing process used to form a DS3 signal.

22 What method is used to provide a 1s density when transmitting a DS2 signal?

23 Describe the M13 framing format. What is the function of the C bits in each subframe?

24 What are the functions of the C bits in the DS3 C-bit parity frame format?

25 What is a common application for the use of a DS3 fan-out network structure?

7

T- AND E-CARRIER MULTIPLEXERS AND ACCESS CONCENTRATORS

The key to the effective utilization of T- and E-carrier transmission facilities is multiplexers and access concentrators. As previously noted in Chapter 2, telephone company channel banks contain time division multiplexers. Other types of multiplexing equipment that can be effectively used with T-carrier facilities are terminal or end-unit multiplexers, nodal switches and access concentrators. In this chapter we will examine the operation and utilization of each category of T- and E-carrier multiplexing equipment and access concentrators, focusing our attention on how they can be used to support a variety of end-user transmission requirements.

7.1 CHANNEL BANKS

The channel bank represents the earliest type of multiplexing equipment used to interface a T- and E-carrier transmission facility. Channel banks are used by telephone companies to multiplex digitized voice channels onto a T- or E-carrier transmission facility.

The channel bank was developed by Bell Laboratories and has been used since the 1960s in telephone company central offices with a digital access cross connect (DACS) system to switch digitized voice calls.

Evolution

In the United States, the first set of digital channel banks was known as D1 systems, where the D stands for digital, which was a significant difference in multiplexing in comparison to the prior method of using frequency division multiplexing. The D1 channel bank used a seven-bit PCM algorithm for digitizing voice, and the eighth bit of each eight-bit channel time slot was used for signaling. The D1 channel bank multiplexed the eight bits of each of the 24 channels into a multiplexing frame and added a framing bit to maintain synchronization between channel banks. The framing bit was used to generate a repeating '1010' pattern which is referred to as D1 framing.

The use of an alternating pulse in bit position 193 in the DS1 frame was selected by AT&T because it is simple to generate. This method of bit framing provides a backward-acting method of synchronization. That is, the receiving terminal checks to see if the framing bit is what it expects at the end of the frame. If not, framing is lost and it can take several frames to resynchronize due to the receiving equipment having to check each position in the frame. When an out-of-frame (OOF) condition occurs, between 0.4 and 0.6 milliseconds may be required for its detection. The receiving equipment then attempts resynchronization by using the previous frame's frame bit. The receiving equipment uses a shift register and moves forward one bit position whenever the alternating pattern is violated. Once it successfully finds eight frames, the receiver assumes it is resynchronized. Typically this process requires a worst-case scenario of approximately 50 milliseconds to test all 192 positions.

The sampled channels under D1 framing appeared in the time division multiplexed signal in the following order: 1, 13, 2, 14, 3, 15, 4, 16, ... Figure 7.1 illustrates the D1 channel bank channel number sequence assignment.

The next evolution in the development of carrier multiplexing equipment was the D2 channel bank. D2 channel banks were

Time Slot

1	2	3	4	5	6	7	8	9	10	11	12	13	14	15	16	17	18	19	20	21	22	23	24
1	13	2	14	3	15	4	16	5	17	6	18	7	19	8	20	9	21	10	22	11	23	12	24

Figure 7.1 D1 channel bank channel number sequence assignment

Time
Slot   1   2 3   4   5   6   7   8   9 10 11 12 13 14 15 16 17 18 19 20 21 22 23 24

| 12 | 13 | 1 | 17 | 5 | 21 | 9 | 15 | 3 | 19 | 7 | 23 | 11 | 14 | 2 | 18 | 6 | 22 | 10 | 16 | 4 | 20 | 8 | 24 |

Channel
Assignment

Figure 7.2 D2 channel number sequence assignment. D2 channel banks employ a pseudo-random assignment of the 24 DS0 channels into the 24 time slot positions of the frame

deployed during 1969 as a result of D1 channel banks not providing an adequate signal to noise ratio. When developing the D2 channel bank AT&T also changed its multiplexing method by enabling 96 channels to be multiplexed into four independent DS1 signals. To achieve economies of scale, a single coder and compander was used for all 96 channels organized into eight groups of 12 channels. Each DS0 channel in a group is sampled sequentially and the samples from each group are interleaved to form a 96-channel signal. Once this is accomplished, four DS1 signals are created, each signal containing samples from two of the original eight groups of 12 channels. Figure 7.2 illustrates the channel number sequence assignment used by D2 channel banks.

As previously discussed in Chapter 6, the Superframe Format, commonly referred to as D4 framing, was introduced with the introduction of D2 channel banks. The D2 channel bank also extended the digitization of voice into an eight-bit PCM word and introduced the bit robbing technique discussed in Chapter 6 to obtain a signaling capability to pass such channel signaling information as on-hook, off-hook, and busy.

In 1972 AT&T introduced its D3 channel bank. The D3 channel bank is essentially the same as the D2 channel bank but changed the time slot sequencing in the frame to correspond to the channel numbering sequence. Included in this channel bank was circuitry that provided a minimum 1s density as discussed in Chapter 6. Instead of grouping 96 DS0s like the D2 channel bank, the D3 channel bank combined 24 channels into a single DS1 signal. This resulted in the multiplexing of channels in their sequential order.

The D4 channel bank was placed into service during 1977 and is still used today. This channel bank multiplexes two sets of 24 channels onto two T1 circuits or it can be used to interleave two T1 data streams onto a T1C circuit operating at 3.152 Mbps. Due to the use of bit robbing, data rates are restricted to a maximum of 56 kbps on each DS0 channel of a D4 channel bank.

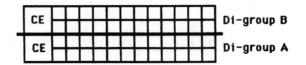

Figure 7.3 D4 channel bank: CE=Common equipment

Each D4 channel bank requires the installation of specific circuit cards to provide an interface to a specific type of voice or data circuit. As an example, a 56 kbps office channel unit (OCU) and a 56 kbps DSU are required to interface a 56 kbps data circuit onto a DS0 channel, while a four-wire E&M (ear and mouth) circuit card must be used to interface the channel bank to a four-wire E&M circuit.

The D4 channel bank is divided into two groups called di-groups, with each group containing 24 channels as illustrated in Figure 7.3. The functions of the common equipment for each di-group are listed in Table 7.1.

As previously discussed in Chapter 6, the introduction of the D4 channel bank continued the extension of framing to 12 frames, a technique referred to as the Superframe framing format. Since D4

Table 7.1 D4 channel bank functional units.

Unit	Operational function
Transmit unit	Performs the function of directing channel sampling encoding and insertion of framing pulses for the di-group
Receive unit	Performs the function of decoding the PCM signal for a di-group, demultiplexing the channel information and extracting the timing, framing and signaling information
Trunk processing unit	Automatically disconnects customers from a circuit during a carrier failure, making the circuits busy so customers cannot seize them, as well as stops all charges on the calls
Power distribution unit	Distributes power to the channel bank
Power control unit	Functions as a dc-to-dc converter to supply correct voltage to the channel bank
Office interface unit	Provides or derives one of three types of clocking, including:
Local timing	Timing is developed within the channel bank and is independent of any other source
Looped timing	Timing is derived from the received signal and is then used for the transmitted signal
External timing	Timing is obtained from an external device

a. Interconnecting two D-Banks.

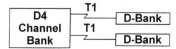

b. Directly connecting two D4 channel banks via a T1C transmission facility.

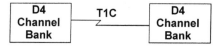

c. Indirectly connecting two D-Banks via a T1C line and M1C multiplexer.

Figure 7.4 The D4 channel bank can operate in several modes

channel banks multiplex two sets of 24 channels, they can operate in several modes.

Figure 7.4 illustrates three common modes of operation for a D4 channel bank. In Figure 7.4a the D4 channel bank uses the output from each di-group as a mechanism to provide 24 channel connectivity with a distant D-Bank over a T1 circuit. Here the term D-Bank is used to reference a 24-channel channel bank or a di-group on a D4 channel bank. In Figure 7.4b two D4 channel banks are shown directly connected to one another via a T1C circuit, providing 48 channels between locations. In Figure 7.4c the D4 channel bank is shown providing indirect communications with two remote D-Banks via the use of an AT&T M1C multiplexer. The M1C multiplexer was developed to convert two DS1 signals to a SD1C signal for transmission via a T1C transmission facility.

The D5 channel bank was introduced by AT&T during 1985. Very similar to the D4 channel bank, the D5 channel bank performs the same basic functions to include the use of the Superframe. Unlike the D4 channel bank that supports only 24 or 48 channel multiplexing, the D5 channel bank added support for multiplexing 72 and 96 DS0s.

Recognizing the fact that it was difficult to detect errors by simply examining the framing bit, AT&T introduced the Extended Super-frame Format (ESF) as an option with the D5 channel bank. As previously described in Chapter 6, the ESF consists of a group of 24 DS1 frames, with six framing bits used for synchronization, six used to store a six-bit Cyclic Redundancy Check, and 12 used to create a 4 kbps data link. Through the use of the CRC it became possible to monitor the performance of the circuit without having to take it out of service. This type of testing is referred to as non-intrusive testing.

Channel banks versus T- and E-carrier multiplexers

The major differences between channel banks and T- and E-carrier multiplexers are in the areas of voice interfaces, diagnostic capability, and the ability to perform automatic rerouting of data. Although channel banks support a large variety of voice interfaces, they have limited diagnostics and cannot be used for rerouting by an end-user. In comparison, T- and E-carrier multiplexers may have a limited voice interface capability; however, they usually have superior diagnostics and many provide the capability to automatically reroute data.

Special carrier multiplexing facilities

AT&T currently offers two special types of multiplexing service marketed under the names M24 and M44. Since equivalent services are offered by other communications carriers, we will review their functionality in this section.

M24 multiplexing

M24 multiplexing is a service provided under Accunet T1.5 which allows a T1 line to be broken out into 24 individual lines. The resulting individual lines can access the PSTN, DDS service or leased lines as illustrated in Figure 7.5. Note that under AT&T's M24 service, a digital access and cross connect system (DACS) is used to route specific DS0 channels to a variety of other carrier facilities, as indicated in the figure.

M44 multiplexing

M44 multiplexing is a service which compresses voice signals from two T1 lines onto one T1 line and expands compressed voice signals back into two T1 lines.

The key to M44 multiplexing is the use of adaptive differential pulse code modulation (ADPCM) at the carrier's central office where M44 multiplexing is performed. At that location 22 DS0 channels on each of two T1 lines are first demultiplexed by a DACS. The resulting 64 kbps channels are passed to an M44 multiplexer which encodes each channel according to the ADPCM algorithm and returns the data to the DACS which then bundles 44 DS0 channels onto a T1 circuit. Figure 7.6 illustrates the M44 multiplexing process.

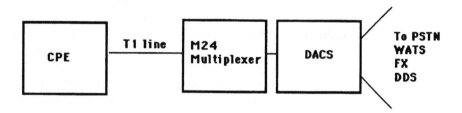

Figure 7.5 M24 multiplexing. M24 multiplexing permits a customer to access a wide assortment of carrier facilities via a common T1 line: CPE = customer premises equipment, DACS = digital access cross connect system, PSTN = public switched telephone network, FX = foreign exchange, DDS = Dataphone Digital Service

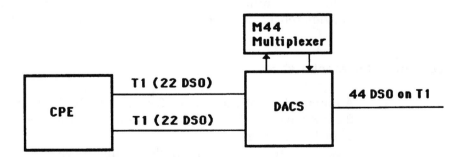

Figure 7.6 M44 multiplexing. M44 multiplexing combines 44 DS0 channels onto one T1 circuit through the use of adaptive differential pulse code modulation

M44 service can be used between central offices or between a central office and a customer's premises. Although this service offers a significant economic improvement over the use of PCM encoding, it is limited to compressing only voice, requiring data to be passed through uncompressed.

7.2 T- AND E-CARRIER MULTIPLEXERS

A T- or E-carrier multiplexer is a device that can be used to integrate voice, data and video onto a T- or E-carrier transmission facility.

T1 facilities were first offered to the public as a tariffed service by AT&T in 1983. The availability of AT&T's Accunet T1.5 service was a driving force in a literal explosion of new and established companies introducing products for use with this service.

T- and E-carrier multiplexers were originally point-to-point devices with each multiplexer considered to be a terminal or end-unit device that provided neither networking capability nor the ability to dynamically assign bandwidth utilization. Commensurate with the growth in T- and E-carrier networking has been a corresponding increase in the features and capabilities of T- and E-carrier multiplexers. Many T- and E-carrier multiplexers now include multinodal support capability and the ability to perform many types of voice digitization through the addition of voice digitization modules, as well as the ability to dynamically assign voice, data and video to T- and E-carrier bandwidth.

Since the primary difference between a T- and E-carrier terminal or end-unit multiplexer and a T- and E-carrier nodal switch is in the areas of multi-trunk support capability and the automatic rerouting of data, we will first focus our attention upon their common operational characteristics and features in this section. Using this information as a base will then enable us to examine the key differences between these two types of multiplexers.

Operational characteristics

A minimally configured T- or E-carrier multiplexer supports from 24 or 30 to several hundred DS0 inputs and from one to 12 or more T- or E-carrier lines.

Some T- and E-carrier multiplexers digitize voice directly through the addition of optional voice digitizer modules. Other multiplexers require digitized voice to be routed as input to the multiplexer. Figure 7.7 illustrates a typical T- or E-carrier multiplexer

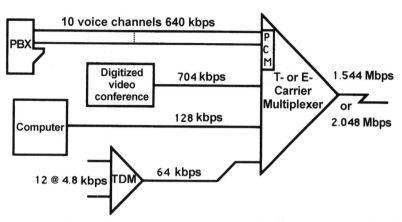

Figure 7.7   Typical T1 or E1 multiplexer application

application where the device is used to combine a variety of digitized voice, data and video inputs onto a T-carrier facility operating at 1.544 Mbps (North American) or an E1-carrier operating at a 2.048 Mbps data rate.

In the example illustrated in Figure 7.7, it was assumed that the digitized videoconferencing required the use of 11 DS0 channels. Hence, the operating rate required to support full-motion video through the multiplexer becomes 64 kbps times 11, or 704 kbps. The 10 lines routed from the PBX are assumed to be analog, resulting in a requirement for voice digitizer modules to be installed in the multiplexer. In this example, PCM digitization modules were used, resulting in each analog channel digitized at 64 kbps onto one DS0 channel. Thus, the 10 voice channels are shown as collectively occupying 640 kbps of bandwidth.

In the lower portion of Figure 7.7, 12 4.8-kbps data sources are first multiplexed by a conventional TDM, resulting in the composite 57.6 kbps data stream boosted to 64 kbps by the use of pad bits to enable its support by one 64 kbps channel. The use of this type of multiplexer to pre-multiplex low data rate asynchronous or synchronous data sources will depend upon the functionality of the T- or E-carrier multiplexer and its ability to multiplex subrate digital data streams. Some multiplexers are limited to multiplexing synchronous data only. Other multiplexers may support a variety of asynchronous and synchronous data rates through the use of different channel cards. Table 7.2 lists some of the more common data rates supported by T- and E-carrier multiplexers, including the operating rate of many voice digitization modules.

Table 7.2 Typical T1 and E1 multiplexer channel rates.

Type	Data rates (bps)
Asynchronous	110, 300, 600, 1200, 1800, 2400, 3600, 4800, 7200, 9600, 19 200
Synchronous	2400, 4800, 7200, 9600, 14 400, 16 000, 19 200, 32 000, 38 400, 40 800, 48 000, 50 000, 56 000, 64 000, 112 000, 115 200, 128 000, 230 400, 256 000, 460 800
Voice	8000, 16 000, 32 000, 48 000, 64 000

Multiplexing efficiency

In addition to examining the type of voice, data and video support, it is also important to determine how efficiently a T- or E-carrier multiplexer utilizes each DS0 channel. Some multiplexers can place only one asynchronous or synchronous data source onto a DS0 channel regardless of its data rate, whereas other multiplexers may make much more efficient utilization of DS0 channels. In our examination of T- and E-carrier multiplexer features, which follows this section, we will investigate the advantages of a subrate multiplexing capability.

Features to consider

Table 7.3 contains a list of the major features of T-carrier multiplexers that warrant consideration during an acquisition process. While all of the listed features are important to consider, they may not be relevant to certain situations based upon the immediate and long-term requirements of a specific organization.

Bandwidth utilization

Inefficient T- and E-carrier multiplexers assign data to the composite transmission facility using 64 kbps DS0 channels for each data source as illustrated in Figure 7.8. In this example the assignment of input data sources is fixed to predefined channels, resulting in the inability of the multiplexer to take advantage of the inactivity of different data sources.

Table 7.3 T- and E-carrier multiplexer features to consider.

Bandwidth utilization method
Bandwidth allocation method
Voice interface support
Voice digitization support
Internodal trunk support
Subrate channel utilization
Digital access cross connect capability
Gateway operation support
Alternative routing and route generation
Redundancy
Maximum number of hops and nodes supported
Diagnostics
Configuration rules

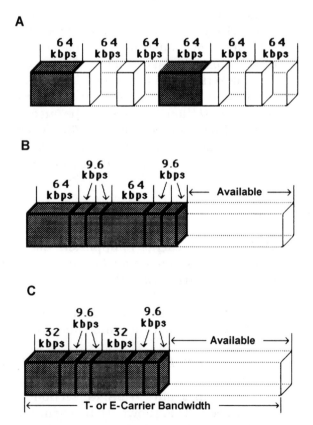

Figure 7.8 Demand assigned bandwidth: A conventional bandwidth allocation with PCM digitization; B demand assigned bandwidth with PCM digitization; C demand assigned bandwidth with ADPCM digitization

More efficient multiplexers employ a variety of demand assigned bandwidth techniques to make more efficient use of the composite T- or E-carrier bandwidth. This is illustrated in Figure 7.8 in which a basic demand assignment feature of a T- or E-carrier multiplexer dynamically assigns bandwidth based upon the activity of the data sources. In this example it was assumed that several 9.6 kbps data sources became active along with two PCM digitized voice conversations and were dynamically assigned to the bandwidth in their order of activation. Note that this method of bandwidth assignment normally results in an increase in available bandwidth since the probability of all inputs becoming active at one time is usually very low. In addition, having the capability to allocate bandwidth based upon the data rate of the data sources and not on a DS0 channel basis allows the 9.6 kbps data sources to occupy significantly less bandwidth. Thus, demand assignment with dynamic bandwidth allocation results in a considerable improvement in the use of a T- or E-carrier's data transmission capacity in comparison to a conventional bandwidth allocation process.

Figure 7.8C illustrates the effect upon bandwidth allocation based upon the use of a more efficient voice digitization module in multiplexers. In this example it was assumed that ADPCM voice digitization modules were used in the multiplexer instead of PCM voice digitization modules. The use of ADPCM reduces the bandwidth required for carrying each voice conversation to 32 kbps, further increasing the available bandwidth of the T- or E-carrier to support other data sources.

Bandwidth allocation

Most T- and E-carrier multiplexers use time division multiplexing schemes to allocate bandwidth to each voice and data channel as well as portions of DS0 channels. Techniques used for bandwidth allocation can include the demand assignment of bandwidth previously illustrated in Figure 7.8B and C, as well as non-contiguous resource allocation and the packetization of voice, data and video.

Figure 7.9 illustrates the advantage of non-contiguous resource allocation over the conventional method of allocating DS0 channels. In Figure 7.9A, a section of bandwidth supporting three voice calls is shown. Here, each call is placed in a contiguous portion of the bandwidth. In Figure 7.9B it was assumed that call B was completed and its bandwidth became available for use. Now suppose a data source (D) became active that required more

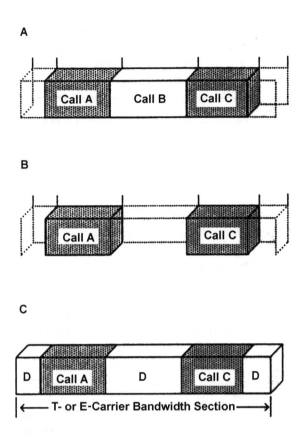

Figure 7.9 Bandwidth allocation methods. Non-contiguous resource allocation of bandwidth enables input to the T- or E-carrier multiplexer to be split into portions of the available bandwidth: A T- or E-carrier bandwidth section supporting three calls; B call B is completed and its bandwidth becomes available; C data source D multiplexed into non-contiguous sections of bandwidth

bandwidth than that freed by the completion of call B. Under the non-contiguous resource allocation method the bandwidth required to accommodate the data source could be split into two or more non-contiguous sections of the T- or E-carrier bandwidth as illustrated in Figure 7.9C.

The third method of bandwidth assignment was also pioneered by Stratacom with their introduction of a T-carrier multiplexer that packetizes both voice and data. The Stratacom multiplexer uses 'fast packet' technology where the term 'fast packet' refers to the fact that information is transmitted across the network in packet format instead of in a time division multiplexed format. Although the external interface to voice and data is the same as in a

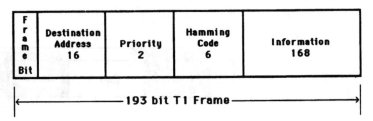

Frame Bit	Destination Address 16	Priority 2	Hamming Code 6	Information 168

|← ——————————— 193 bit T1 Frame ——————————— →|

Figure 7.10 Stratacom packet format. The Stratacom T1 multiplexer packetizes voice and data sources into 193 bit frames containing 168 information bits

conventional T-carrier multiplexer, the internal operation of the Stratacom multiplexer is considerably different from other devices.

Another interesting method of bandwidth allocation involves the packetization of voice and data sources onto T- and E-carrier facilities. This technique was also pioneered by Stratacom.

The Stratacom fast packet multiplexer generates packets only when data sources are active, using a packet length of 193 bits which corresponds to the North American T1 frame length. Figure 7.10 illustrates the Stratacom frame format. Although 20 bits, in effect, function as overhead to provide a destination address (16 bits), priority (two bits), and error correction to the header by the use of a hamming code (six bits), the efficiency of packetized multiplexing can be considerable. This is because the technique takes advantage of the fact that voice conversations have periods of silence and are typically half-duplex in nature. This enables packet technology to provide an efficiency improvement of approximately 2 : 1 over conventional time division multiplexing of voice. With the addition of ADPCM voice digitization modules, the Stratacom fast packet multiplexer can support up to 96 voice conversations on a T1 circuit.

One of the problems associated with the use of packet technology to transport digitized voice is the fact that you cannot delay voice. Thus, unlike data packets that can be retransmitted if an error is detected, packetized voice cannot tolerate the delay of retransmission. In addition to not being able to retransmit voice, you must also consider the effect of a large number of voice channels becoming active. When too many channels become active, the total bit rate of the digitized input channels can exceed the output bit rate of a T- or E-carrier circuit. To avoid too much delay to specific channels, some channels will be skipped since a listener can tolerate a 125-millisecond delay. Another problem associated with packetized voice is the delay that can occur as the packets are routed from node to node in a complex network. To overcome this problem, Stratacom incorporates a priority field in its packet which enables certain packets to be processed and routed before other types of packets.

Voice interface support

Since most T- and E-carrier multiplexer applications include the concentration of voice signals, the type of voice interfaces supported for two-wire and four-wire applications is an important multiplexer feature to consider. Prior to examining the types of voice interfaces supported by T- and E-carrier multiplexers, let us first review some of the more common types of voice signaling methods since it is the signaling method that is actually supported by a particular interface.

Two of the most common types of telephone signaling include loop signaling and E&M signaling. Loop signaling is a signaling method employed on two-wire circuits between a telephone and a PBX or between a telephone and a central office. E&M signaling is a signaling method employed on both two-wire and four-wire circuits routed between telephone company switches.

In loop signaling, the raising of the telephone handset results in the activation of a relay at the PBX or central office, causing current to flow in a circuit formed between the telephone set and the PBX or central office. The raising of the handset, referred to as an off-hook condition, results in the PBX or central office returning a dial tone to the telephone set. As the subscriber dials the telephone number of the called party, the dialed digits are received at a telephone company central office which then signals the called party by sending signaling information through the telephone company network. Once the call is completed the placement of the handset back onto the telephone set, a condition known as on-hook, causes the relay to be deactivated and the circuit previously formed to open.

A second type of telephone off-hook signaling that flows in a loop is ground start signaling. This method of signaling is also used on two-wire circuits between a telephone set and a PBX or central office. Unlike loop start signaling, in which loop seizure is detected at the PBX or central office, ground start allows the detection of loop seizure to occur from either end of the line.

E&M signaling

E&M signaling is used on both two-wire and four-wire circuits connecting telephone company switches. Here the M lead is used to send ground or battery signals to the signaling circuits at a telephone company switch, while the E lead is used to receive an open or ground from the signaling circuit. In E&M signaling the local end asserts the M lead to seize control of the circuit. The remote end receives the signal on the E lead and toggles its M lead as a signal for the local end to proceed. The local end then sends the address by toggling its M lead, in effect, placing dialing pulses on that lead which is used by the remote end to effect the desired connection. Once a call is completed, either party will drop its M lead, resulting in the other side responding by dropping its M lead.

There are five types of E&M signaling referred to as Types I through V, while a sixth popular method is a British Telecom standard used in the United Kingdom. The primary difference between E&M signaling types relates to the method by which an on-hook condition is established (ground or open) and the device that supplies battery (PBX and/or switch).

Under E&M Type I signaling, the battery for both E&M leads is supplied by the PBX. At the PBX, an on-hook condition results in the M lead being grounded and the E lead open. In comparison, an off-hook condition results in the M lead providing the battery and the E lead being grounded. Type I signaling is the most commonly used four-wire trunk interface in North America.

One of the problems associated with Type I signaling is the fact that the interface can result in a high return current through the grounding system. For example, if two PBXs were improperly grounded this situation could result in current flowing down the M lead, resulting in a remote PBX detecting current on the E lead and causing a false seizure of a trunk. E&M Type II signaling addresses this problem by addressing two additional signaling leads— battery (SB) and signal ground (SG). Under Type II signaling the E lead operates in conjunction with the SG lead while the M lead is strapped to the SB lead, grounding the trunk at the end and eliminating potential rounding problems.

Type III signaling is similar to Type I, with the main difference being in the use of transmission equipment to supply the battery and grounding source. Type II signaling was primarily used with older central office equipment and is in limited use due to the replacement of most older central office switches.

Type IV signaling is similar to Type II, with the key difference in the operation of the M lead. In Type II signaling the M leads are 'open' and 'battery'. Under Type IV signaling the states are 'ground' and 'open' which prevent an accidental shorting of the SB lead, resulting in an excessive current flow.

Table 7.4 Typical T- and E-carrier voice interface modules.

Two-wire transmission only
Two-wire E&M
Two-wire foreign exchange
Four-wire transmission only
Four-wire E&M

The most popular method of E&M signaling outside North America is Type V, an ITU standard. Under Type V signaling both the switch and transmission equipment supply a battery. The battery for the M lead is located in the signaling equipment while the battery for the E lead is located in the PBX.

Table 7.4 lists some of the more common types of voice interface cards supported by many T- and E-carrier multiplexer vendors. The two-wire and four-wire transmission only interfaces are designed to support permanent two-wire and four-wire connections between two points that do not require the passing of signaling information. Both types of interfaces are normally used to support a data modem connection through a T- or E-carrier multiplexer.

The two-wire and four-wire E&M interfaces usually support the connection of PBXs and telephone company equipment to the multiplexer. As previously mentioned, there are five types of E&M signaling, with each type applicable to both two-wire and four-wire operations.

The two-wire foreign exchange office interface is designed to support the attachment of a multiplexer to a PBX or central office switching equipment that provides an open or closed foreign exchange termination point.

Voice digitization support

In addition to the method of bandwidth assignment and allocation a third major feature affecting the efficiency of a T- or E-carrier multiplexer is the type of voice digitization modules the device supports. Although most multiplexers support the use of PCM and ADPCM, some vendors also support the use of adapter cards that contain proprietary voice digitization modules. One example of this is adaptive speech interpolation which changes the digitization rate of selected voice channels from 32 kbps to 24 kbps as available bandwidth becomes saturated. Although some proprietary techniques may offer advantages both in the fidelity of a reconstructed

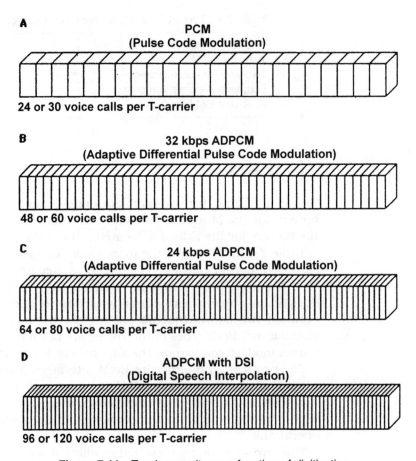

A

PCM
(Pulse Code Modulation)

24 or 30 voice calls per T-carrier

B

32 kbps ADPCM
(Adaptive Differential Pulse Code Modulation)

48 or 60 voice calls per T-carrier

C

24 kbps ADPCM
(Adaptive Differential Pulse Code Modulation)

64 or 80 voice calls per T-carrier

D

ADPCM with DSI
(Digital Speech Interpolation)

96 or 120 voice calls per T-carrier

Figure 7.11 Trunk capacity as a function of digitization

voice signal as well as in the bandwidth required to carry the signal, their use restricts an organization to one vendor's product.

Figure 7.11 illustrates the effect of the use of several types of voice digitization modules upon the capacity of a T1 or E1 circuit. If standard PCM modules are used, the T- or E-carrier becomes capable of supporting either 24 or 30 voice calls depending upon whether a North American or European facility is used. When 32 kbps ADPCM modules are used to digitize voice, the voice carrying capacity of the T-carrier is doubled as shown in Figure 7.11B. Figure 7.11C, which illustrates the use of 24 kbps ADPCM, shows the voice carrying capacity of a T-carrier tripling, while Figure 7.11D shows how the voice carrying capacity of a T-carrier can be quadrupled through the use of ADPCM and DSI.

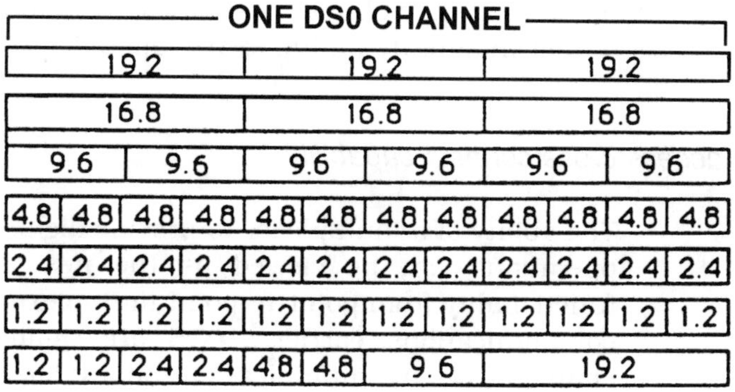

ONE DS0 CHANNEL											
19.2				19.2				19.2			
16.8				16.8				16.8			
9.6		9.6		9.6		9.6		9.6		9.6	
4.8	4.8	4.8	4.8	4.8	4.8	4.8	4.8	4.8	4.8	4.8	4.8
2.4	2.4	2.4	2.4	2.4	2.4	2.4	2.4	2.4	2.4	2.4	2.4
1.2	1.2	1.2	1.2	1.2	1.2	1.2	1.2	1.2	1.2	1.2	1.2
1.2	1.2	2.4	2.4	4.8	4.8	9.6		19.2			

Figure 7.12 Typical subrate channel utilization

Internodal trunk support

The internodal trunk support feature of T- and E-carrier multiplexers refers to the ability of the device to connect to North American and European T- and E-carrier facilities. For example, to support a North American T1 carrier facility usage, the multiplexer must not only operate at 1.544 Mbps but, in addition, support the required communications carrier framing-D4 or ESF. To support a European E1-carrier facility usage the multiplexer must operate at 2.048 Mbps and support CEPT PCM-30 framing.

Subrate channel utilization

Channel utilization is a function of the subrate multiplexing capabilities of the multiplexer. Many T- and E-carrier multiplexers support asynchronous data rates from 50 bps to 19.2 kbps and synchronous data rates from 2.4 kbps to 19.2 kbps, permitting multiple data sources to be placed onto one DS0 channel. Figure 7.12 illustrates one of the methods by which a T- or E-carrier multiplexer vendor's equipment might multiplex subrate data channels onto one DS0 channel. Unfortunately, not all vendors provide a subrate multiplexing capability in their equipment. When this occurs, subrate data sources are bit padded to operate at 64 kbps which can considerably reduce the ability of the multiplexer to maximize bandwidth utilization. In such situations users can obtain one or more subrate multiplexers to combine several data sources to a 64 kbps data rate; however, this may result in a higher cost than obtaining T- or E-carrier multiplexers that include a built-in subrate multiplexing capability.

Digital access cross connect capability

The ability of a multiplexer to provide digital access cross connect operations can be viewed as the next step up in terms of functionality from point-to-point multiplexer operations. Although communications carrier DACS are limited to switching DS0 channels, many multiplexer vendors include the capability to drop and insert/bypass subrate channels or digitized voice encoded at bit rates under 64 kbps, permitting sophisticated T- and E-carrier networks to be constructed.

Figure 7.13 illustrates an example of the use of a digital access cross connect feature used in three T-carrier multiplexers labeled A, B and C. In this example, channel 8 on multiplexer A is routed to multiplexer C where it is dropped, freeing that DS0 channel for use, as the T-carrier is then routed to multiplexer B's location. Thus, a data source at location C could be inserted into the T-carrier on channel 8, resulting in channel 8 being routed from C to B as shown in Figure 7.13. In this example it was assumed that all other DS0 channels were simply passed through or bypassed multiplexer C and were then routed to multiplexer B. Thus, a digital access and cross connect capability can be used to establish a virtual circuit through an intermediate multiplexer (bypass) without demultiplexing the data, to allow intermediate nodes to add data to the data stream (insert), as well as to permit an intermediate node to act as a terminating node (drop) for other multiplexers.

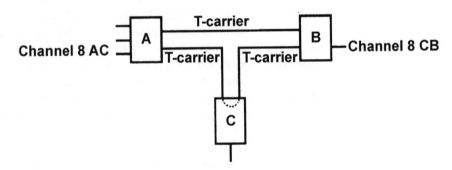

Figure 7.13 DS0 cross connect. Although many multiplexers support the cross connection of DS0 channels, some T- and E-carrier multiplexers also permit the cross connection of subchannels

Gateway operation support

To function as a gateway requires a T- or E-carrier multiplexer to support a minimum of two high speed circuits. In addition, the multiplexer must perform several other operations that must be coordinated with the use of T- or E-carrier multiplexers connected to the gateway multiplexer.

Three of the main problems associated with connecting European and North American circuits through the use of a gateway multiplexer involve compensating for the differences between European and North American transmission facilities with respect to the number of DS0 channels each carrier supports, the method by which signaling is carried in each channel, and the method by which performance monitoring is accomplished.

Since a European E1 carrier contains 30 DS0 channels, while a North American T1 link supports 24, the gateway multiplexer will map 30 DS0 channels to 24, resulting in the loss of six channels. This means that the end-user's organization is limited to the effective use of 24 channels on a European connection via a gateway multiplexer.

For signaling conversion the gateway multiplexer will move AB or ABCD signaling under D4 and ESF frame formats into channel 16 for North American to European conversion. For signaling conversion in the other direction, appropriate bits will be moved from channel 16 to the robbed bit positions used in D4 and ESF framing.

The area of performance monitoring is presently either ignored by gateway multiplexers or handled on an individual link basis to the gateway device. Since European systems use a CRC-4 check while ESF employs a CRC-6 check, no conversion is performed by multiplexers that support performance monitoring, since the results would be meaningless for a link consisting of both North American and European facilities connected through a gateway. Instead, the gateway will provide statistics treating each connection as a separate transmission facility. Figure 7.14 illustrates the placement of a gateway T-carrier multiplexer connecting a North American T1 facility to a European CEPT 30 facility.

Alternate routing and route generation

If a T- or E-carrier network consists of three or more multiplexers interconnected by those carrier facilities, both the alternate routing

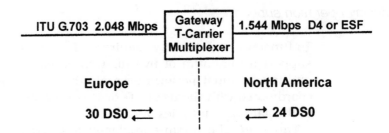

Figure 7.14 Gateway operation. The operation of a gateway results in the dropping of six DS0 channels when a European CEPT 30 facility is converted to a North American T1 link

capability and the method of route generation are important features to consider.

Basically, route generation falls into two broad areas: paths initiated by tables constructed by operators and dynamic paths automatically generated and maintained by the multiplexers. To illustrate both alternate routing and route generation, consider the T-carrier network illustrated in Figure 7.15. In this illustration the T-carrier circuit connecting multiplexers A and B has failed. With an alternate routing capability some or all DS0 channels previously carried by circuit AB must be routed from path AC to path CB to multiplexer B.

If the multiplexers employ alternate routing based upon predefined tables assigned by operations personnel, DS0 channels previously routed on path AB will be inserted into the T-carrier linking A to C and then routed onto path CB based upon the use of those tables. As the DS0 channels from path AB are inserted into the T-carrier linking A to C, DS0 channels on path AC must be

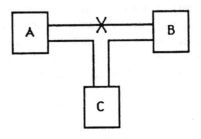

Figure 7.15 Alternate routing. T- and E-carrier multiplexers provide a variety of methods to effect alternate routing to include using predefined tables and by the dynamic examination of current activity at the time of failure

dropped, a process referred to as bumping. If calls were in progress on path A to C and C to B when the failure between A and B occurred, those calls may be dropped, depending upon whether or not the multiplexers employ a priority bumping feature or have the capability to downspeed voice digitized DS0 channels.

Priority bumping refers to the ability to override certain existing DS0 subchannels based upon the priority assigned to DS0 channels that were previously carried on the failed link and the priorities assigned to DS0 active channels on the operational links. Downspeed refers to the capability of multiplexers to shift to a different and more efficient voice digitization algorithm to obtain additional bandwidth with the ability to carry DS0 channels from the failed link on the operational circuits. One example of downspeed would be switching from 32 kbps ADPCM to 24 kbps ADPCM, resulting in freeing up 12 kbps per operational DS0 channel.

When alternate routing and route generation is dynamically performed by multiplexers, those devices examine current DS0 activity and establish alternate routing based upon predefined priorities and the current activity of DS0 channels. If predefined tables are used, the multiplexers do not examine whether or not a particular DS0 channel is active prior to performing alternate routing; however, some multiplexers may have the ability to perform forced or transparent bumping regardless of whether they use fixed tables or dynamically generate alternate paths. Under forced bumping, DS0 channels are immediately reassigned, whereas, under transparent bumping, current voice or data sessions are allowed to complete prior to their bandwidth being reassigned.

Redundancy

Since the failure of a T- or E-carrier multiplexer can result in a large number of voice and data circuits becoming inoperative, redundancy can be viewed as a necessity similar to business insurance. To minimize potential downtime, you can consider dual power supplies as well as redundant common logic and spare voice and data adapter cards. Doing so may minimize downtime in the event of a component failure as many multiplexers are designed to enable technicians to easily replace failed components.

Maximum number of hops and nodes supported

As T- or E-carrier multiplexers are interconnected to form a network, each multiplexer can be considered as a network node.

When a DS0 channel is routed through a multiplexer that multiplexer is known as a hop. Thus, the maximum number of hops refers to the maximum number of internodal devices a DS0 channel can traverse to complete an end-to-end connection.

In addition to the maximum number of hops, users must also consider the maximum number of nodes that can be networked together. The maximum number of addressable nodes that can be managed as a single network is normally much greater than the maximum number of hops supported, since the latter is constrained by the delay to voice as DS0 channels are switched and routed through hops.

Diagnostics

Most T- and E-carrier multiplexers provide both local and remote channel loop-back capability to facilitate fault isolation. Some multiplexers have built-in test pattern generation capability which may alleviate the necessity of obtaining additional test equipment for isolating network faults. Refer to Chapter 8 for specific information concerning the testing of digital facilities and the use of built-in and stand-alone test equipment.

Configuration rules

Figure 7.16 illustrates a typical T- or E-carrier multiplexer cabinet layout which is similar to the manner in which a multiplexer would be installed in an industry-standard 19-inch rack. In examining multiplexer configuration rules, a variety of constraints may exist to include the number of trunk module cards, voice cards and data cards that can be installed. Other constraints will include the physical number of channels supported by each card and the type of voice digitization modules that can be obtained. Depending upon end-user requirements, additional expansion shelves may be required to support additional cards. When this occurs, additional power supplies may be required and their cost and space requirements must be considered.

Multiplexers and nodal processors

Due to the liberty by which vendors can label products, there is no definitive line that separates a T- or E-carrier terminal or end-unit

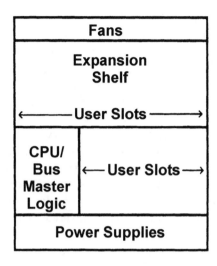

Figure 7.16 Typical T- or E-carrier multiplexer cabinet

multiplexer from a nodal processor. In general, we can categorize each device by the number of trunks they support, the method by which alternate routing is performed, and the method by which operators control the network.

In general, a nodal processor supports more than 16 T- or E-carrier circuits and includes the capability to dynamically perform alternate routing based upon one or more algorithms. In addition, this device normally permits network configuration to be effected from a central node. In comparison, a T- or E-carrier terminal or end-unit multiplexer supports up to 16 trunks and usually relies upon the use of predefined tables to perform alternate routing, assuming they actually have this capability. In addition, network reconfiguration may require operators to reprogram each multiplexer individually.

7.3 ACCESS CONCENTRATORS

In the evolution of communications equipment, LANs, videoconferencing and packet network services, including X.25 and Frame Relay, followed the development of channel banks and T- and E-carrier multiplexers. Recognizing the need of organizations to economize upon the use of multiple access lines to support transmission requirements from one location for a number of different services, vendors began to develop access concentrators during the mid-1990s. In terms of the applications supported, an access concentrator can be considered to represent a very intelligent T- or E-carrier multiplexer. The key difference is the fact that the former is an edge networking device that typically connects to equipment in a carrier's central office. At the carrier's central office the composite data stream is broken out to interface with previously selected services, such as Frame Relay, X.25, fractional T1 or fractional E1, as well as individual DS0s connected to the PSTN. In comparison, T- or E-carrier multiplexers are commonly used in pairs to provide an end-to-end transmission path through the serving carrier's central office.

Another important difference between access concentrators and T- and E-carrier multiplexers concerns their transmission architecture. T- and E-carrier multiplexers are based upon time division multiplexing. Although some access concentrators are also based upon that technology, other access concentrators are based upon ATM technology. Through the use of ATM cell-based transmission which represents a statistical multiplexing technique, the composite line routed to the carrier can be used more efficiently. In addition, a WAN connection of up to 622 Mbps becomes possible, which greatly exceeds the T3 transmission rate of approximately 45 Mbps.

Operation

Figure 7.17 illustrates the potential use of an access concentrator as an edge networking device. In this example the access concentrator statistically multiplexes data from a router connected to a LAN, digitized voice channels from a PBX, and a videoconference facility. The access device would be configured to provide priority under ATM to real-time transmission, such as voice and video. At the service provider's central office the ATM data flow would be broken out to predefined packet network, leased line, and fractional leased line and voice network services. Since the access concentrator requires the use of only a single access line, a considerable saving on the monthly cost of one access line versus multiple access lines becomes possible. Although a single network connection may imply a potential major catastrophe in the event of its failure, in reality many organizations that have a sufficient data transmission requirement to justify an access concentrator will also obtain an FDDI or SONET/SDH ring connection to their location. This will provide a mechanism to recover from failure. The use of optical rings is described in Chapter 8.

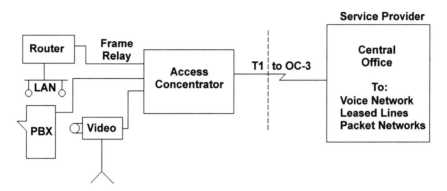

Figure 7.17 Using an access concentrator

Interface capability

The flexibility of an access concentrator depends upon its interface capability. When examining interfaces you should focus upon both end-user and service provider interfaces.

End-user interface

The end-user interface governs the capability of the access concentrator to support different types of data streams. Common interfaces supported by access concentrators include V.35 for some routers and videoconferencing equipment, a High Speed Serial Interface (HSSI) for other routers, and T1 and E1 interfaces for PBXs.

Service provider interfaces

As previously mentioned, access concentrators commonly use ATM cell-based transmission on their high speed transmission facility. Common service provider interfaces range from T1 and E1 through OC-3.

Applications

The major application for access concentrators is to aggregate traffic for connectivity to communications carrier services. Some

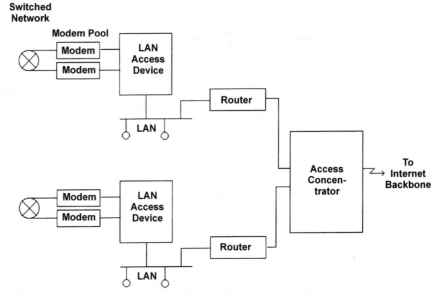

Figure 7.18 Using an access concentrator to support modem users accessing an Internet Service Provider

access concentrators are capable of terminating up to 150 T1 or FT1 circuits, permitting a true fan-out network structure to connect geographically separated locations to a headquarters via a single access line. When used to support Internet Service Providers (ISPs), a single access controller can be used to link modem pools connected to multiple LANs to an Internet Network Service Provider (NSP). Figure 7.18 illustrates the use of an access concentrator by an ISP.

Used in this manner the access concentrator can support several thousand ISP modem users. Thus, the access concentrator represents a versatile networking device that can be used by organizations requiring a significant amount of communications capability.

REVIEW QUESTIONS

1 What is D1 framing?

2 What was the reason for AT&T deploying D2 channel banks?

3 Why is the data rate on a DS0 channel limited to 56 kbps when the channel is routed through a D4 channel bank?

4 Describe three D4 channel bank modes of operation.

5 What is non-intrusive testing?

6 Discuss three differences between channel banks and T- and E-carrier multiplexers.

7 Why would you consider the use of M24 multiplexing or an equivalent service?

8 What is the major benefit obtained from the use of M44 multiplexing or an equivalent service? What is one limitation associated with the use of this service?

9 What are the primary differences between the functionality of T- and E-carrier terminal or end-unit multiplexers and T- and E-carrier nodal switches?

10 Illustrate the advantage obtained by the use of a non-contiguous resource allocation feature of a T- or E-carrier multiplexer.

11 What are two problems associated with the use of applying packet technology to route calls through a series of T- or E-carrier multiplexers?

12 Under what circumstances would you want to use a two-wire or four-wire transmission-only voice interface on a T- or E-carrier multiplexer?

13 What is the most common type of four-wire trunk interface signaling method used in North America?

14 What is the most common type of four-wire trunk signaling method used outside North America?

15 What is the maximum number of voice channels that can be carried on T1 and CEPT 30 facilities if the T-carrier multiplexer uses PCM to digitize voice? How do those numbers change when 32 kbps and 24 kbps ADPCM are used?

16 How could you maximize the efficiency of a T- or E-carrier multiplexer that does not support subrate multiplexing?

17 Discuss three functions a gateway T- or E-carrier multiplexer must perform.

18 Describe two methods used by T- or E-carriers to effect the alternate routing of DS0 channels. Which method would normally be more efficient?

19 What does the term downspeed mean?

20 What does the term transparent bumping refer to?

21 What is the difference between a multiplexer functioning as a hop and a multiplexer functioning as a node?

22 Describe two differences between an access concentrator and a T- or E-carrier multiplexer.

23 How can an access concentrator reduce the cost of networking?

24 What are some common types of access concentrator end-user interfaces?

25 What are some common types of access concentrator service provider interfaces?

8

SONET AND SDH

From a subscriber perspective T- and E-carrier transmission facilities are commonly used to access the Internet and other transmission services as well as to construct private networks. From a communications carrier perspective, the creation of a network based upon asynchronous transmission facilities created a variety of timing problems and the need to employ bit stuffing and other methods to overcome such problems. As the transmission requirements of subscribers continued to grow, communications carriers recognized the necessity to develop a new infrastructure to use the transmission capacity of fiber optic systems that were initially proprietary. This recognition resulted in the development of SONET and SDH, the focus of this chapter.

In this chapter we will primarily focus our examination on SONET. The rationale for this is the similarity between SONET and SDH. However, when appropriate, we will discuss the difference between the two as we review the concepts behind SONET. In addition, we will also focus our attention upon SDH as a separate entity in the second section in this chapter.

8.1 EVOLUTION

Prior to the development of SONET (Synchronous Optical NETwork), the use of fiber optic systems in the public telephone network resulted in proprietary architectures, equipment, line codes, multiplexing formats and maintenance procedures. Commencing during the 1970s, the installation of proprietary fiber optic based systems precluded the ability of communications carriers to mix equipment from different vendors, resulting in the inability of carriers to obtain economy of scale pricing from vendors.

SONET

During 1984 the Exchange Carriers Standards Association (ECSA) initiated an effort to develop a standard to connect one fiber optic system to another. This effort was focused upon achieving three basic objectives:

1 Provide standardized operating rates and formats for transmission over a fiber optic medium.

2 Support the transport of older T- and E-carrier transmission facilities via a synchronized network environment.

3 Specify necessary optical and electrical signals.

The previously described objectives were satisfied by the development of SONET, which was standardized by the American National Standards Institute (ANSI) in 1997.

SDH structure

The effort of ANSI was oriented towards accommodating T-carriers with a DS1 consisting of 24 64-kbps DS0 channels. In Europe, E-carriers were formed based upon 32 64-kbps DS0s used to develop an E1 signal. To accommodate both DS1 and E1 signals, the Consultative Committee for International Telephone and Telegraph (CCITT), which is now known as the ITU, worked on the definition of a standard to provide an internetworking capability between ANSI and CCITT transmission hierarchies. That effort resulted in the Synchronous Digital Hierarchy (SDH) standard which was published in 1989. The primary difference between SONET and SDH involves their basic transmission rates. SONET begins at an operating rate of 51.84 Mbps, while SDH begins at 155.52 Mbps.

8.2 THE SONET STRUCTURAL HIERARCHY

SONET defines a networking technology for transporting signals via a synchronous, optical hierarchy. The actual SONET structural hierarchy can be viewed as consisting of two dimensions. In the horizontal dimension SONET defines the method by which data is transmitted between repeaters and other devices where the optical signal is transmitted and received. In the vertical dimension SONET defines a bit rate hierarchy formed through the use of byte multiplexing. Because an appreciation of the frame structure used by SONET requires knowledge of its line layering and operating rate structure, we will first examine the SONET structural hierarchy in this section.

The line structure

The SONET line structure recognizes the fact that an optical signal can flow through many intermediate devices on its path between source and destination. This results in a line layered structure which provides the ability to subdivide the transmission path into distinct entities that facilitates its operation, administration and maintenance. This subdivision makes it easier to refer to a portion of a path rather than a complete path. In addition, it facilitates performance measurements, testing and troubleshooting on a segment basis which may enable problems to be quickly isolated and corrected.

Figure 8.1 illustrates the line-layered structure of SONET. At the lowest layer, a section refers to transmission between two repeaters or other locations where the optical signal is transmitted and received. At the next higher layer, the line represents a unit of SONET transmission capacity in terms of a Synchronous Transport Signal (STS) whose operating rates are covered next in this section. At the top of the line hierarchy is the path, which represents the end-to-end transmission of an STS payload via SONET.

By subdividing the line into distinct entities it becomes possible to assign operations, administration and overhead functions. Some overhead information permits a channel for technicians to communicate between locations, while other overhead information will transport framing and performance monitoring information. SONET overhead corresponds to the layer structure illustrated in Figure 8.1. That is, overhead is associated with sections, lines and paths.

The SONET transmission hierarchy

Earlier in this book we noted that SONET and SDH provide a hierarchy of digital transmission capacities. The SONET fundamental digital signal is known as the Synchronous Transport Signal level 1 (STS-1) and has an operating rate of 51.84 Mbps. Higher level Synchronous Transport Signals have bit rates that correspond to multiples of the STS-1 signal and are formed by byte-interleaved multiplexing.

STS signals are neither electrical nor optical. Instead, they refer to a sequence of bits. Thus, they more properly represent an abstract signal. For each STS signal there exists a corresponding optical transmission signal (OC). OC is a mnemonic for optical carrier. The OC format is based upon the corresponding STS signal

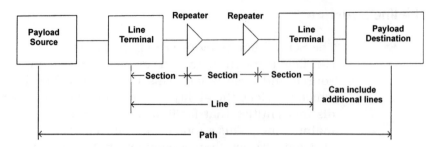

Figure 8.1 SONET's line layer structure

format. The two lowest levels of SONET also have defined electrical signals, referred to as EC-1 and EC-3. Both ECs facilitate the use of test instruments as well as the interconnection of transmission equipment. Table 8.1 summarizes the SONET transmission hierarchy.

The SONET frame structure

SONET uses a basic digital signal referred to as the Synchronous Transport Signal level 1 (STS-1) that operates at 51.84 Mbps. Higher level signals are integral multiples of the STS-1 signal. Thus, an appropriate place to begin an examination of the SONET frame structure is with the structure of the STS-1 signal.

The STS-1 frame format

Figure 8.2 illustrates the frame format of the STS-1 signal in the manner by which most publications depict the serial sequence of SONET STS-1 bytes. That is, they are depicted as a sequence of nine rows, with each row containing 90 bytes. The actual bytes in the overhead section repeat every 90 bytes, enabling a two dimensional illustration of the sequential series of SONET STS-1 bytes. Note that the STS-1 frame is 810 bytes long and repeats at the 8000 frame per second rate of T1 and E1 carriers. Thus, 810 bytes/frame × 8000 frames/s results in a line operating rate of 51.84 Mbps.

In examining the STS-1 frame format, note that nine rows by three columns represents overhead, reducing the payload capacity to 50.112 Mbps. This payload is sufficient to transport a T3 or E3 payload and, as we will shortly note, special overhead bytes referred to as payload pointers indicate where the payload begins

Table 8.1 The SONET digital hierarchy.

Abstract signal	Buildup	Optical signal	Electrical signal	Bit rate (Mbps)
STS-1	N/A	OC-1	EC-1	51.84
STS-3	3XSTS-1	OC-3	EC-3	155.52
STS-12	12XSTS-1	OC-12	N/A	622.08
STS-48	48XSTS-1	OC-48	N/A	2488.32
STS-192	192XSTS-1	OC-192	N/A	9953.28

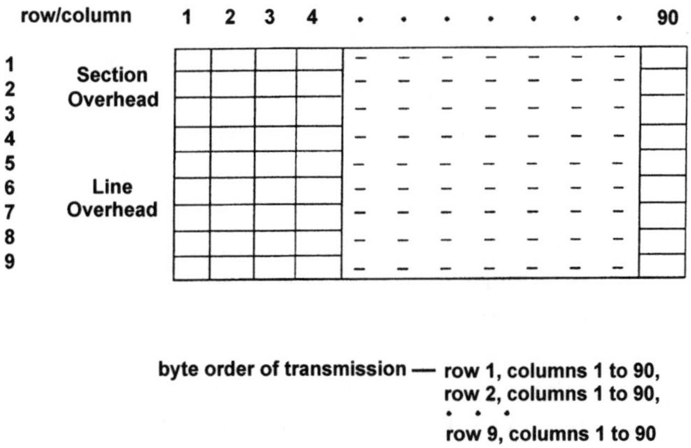

Figure 8.2 The STS-1 frame

within the SONET frame. Such pointers enable individual DS1 frames to be synchronized, resulting in the SONET frame being synchronized with tributary frames and providing the rationale for the 'S' in SONET.

Overhead

As indicated in Figure 8.2, the first three columns of the STS-1 frame provide Section layer and Line layer overhead. The remaining 87 columns, which actually are transmitted as a serial sequence of bytes after each series of overhead bytes, flow one row after another. Since the payload area represents 87 data bytes contained in nine columns, the gross payload is 783 bytes per STS-1 frame.

Payload

Although not indicated in Figure 8.2 the payload area is also subdivided. That subdivision is into two areas, with one for path overhead and one for the actual payload. A total of nine bytes are used for path overhead, reducing the net payload area to 774 bytes.

The gross payload area is organized into a floating position within the STS-1 frame, with its actual position defined by two line overhead bytes—H1 and H2. In SONET terminology the gross payload area, which consists of 783 bytes per frame that floats within the STS-1 frame, is referred to as a synchronous payload envelope (SPE). The floating of the payload represents a mechanism developed to overcome differences in synchronization between different T- and E-carrier transmission facilities.

Figure 8.3 illustrates the positioning of the Synchronous Payload Envelope with respect to a sequence of STS-1 frames. Note that each SPE begins with the first path overhead byte, with the remaining eight path overhead bytes falling within the same column of the STS-1 frame. As indicated in Figure 8.3, the first path overhead byte has the designator J1. We will shortly review the use of each of the path overhead bytes as well as the bytes in the transport overhead area. Since a total of nine bytes from the gross payload area are used for path overhead, this reduces the number of bytes in the frame that are available for transporting an embedded data stream to 774. Thus, the net payload area per STS-1 frame is 774 bytes.

The STS SPE can begin anywhere within the STS-1 envelope. It typically begins in one STS-1 frame and ends in the next as illustrated in Figure 8.3. The two line overhead bytes (H1 and H2) are unsigned integers. They represent an offset in bytes within the gross payload area just after the line overhead byte H3 which defines the position of the SPE in the frame. The path overhead column shown in Figure 8.3 commencing with the designator J1 consists of nine bytes that remain within the payload until they are demultiplexed.

Clocking considerations

The key to the positioning of the SPE within the STS-1 frame is the H1 and H2 pointer bytes and the H3 pointer action byte. The H1 and H2 bytes are simply allocated to a pointer that indicates the offset in bytes between the pointer and the first byte of the STS SPE. When a source clock is fast with respect to the STS-1 clock, it becomes necessary to compensate for the clocking differences by a negative justification operation in which the SPE location is shifted to the left. To accomplish this the H3 byte in the line overhead area acts as a dummy or null byte that can be used. Thus, one payload byte can be squeezed out of the SPE and placed into the H3 byte. To notify the receiver that negative stuffing occurred, special coding is

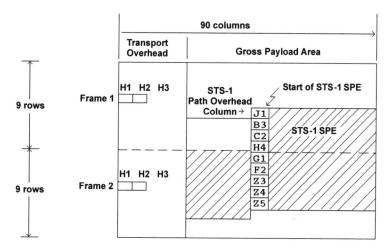

Figure 8.3 Synchronous Payload Envelope (SPE) positioning within an STS-1 frame

placed into byte locations H1 and H2 and the SPE offset is decremented by 1. Figure 8.4 illustrates the use of the H3 byte.

Overhead bytes

In our previous examination of SONET's layer structure we noted that the technology defines a structural hierarchy. That hierarchy begins with sections between repeaters. It is followed by a line which represents a complete STS-N transmission capacity and ends with a path, with the latter representing the end-to-end transmission of an STS-N payload. For each layer—section, line and path—overhead bytes are used to provide an operations, administration, maintenance and provisioning (OAM&P) capability. We will now turn our attention to the use of bytes defined for each SONET overhead area.

Section overhead bytes

Figure 8.5 illustrates the 36 bytes of transport and path overhead associated with an STS-1 payload. Note that rows 1 to 3 of the transport overhead area represent section overhead. Also note that unique alphanumeric designators are assigned to each byte, with the alphabetic prefix used to indicate a general functional category. For example, A bytes are designated for framing while B bytes are designated for bit interleaved parity.

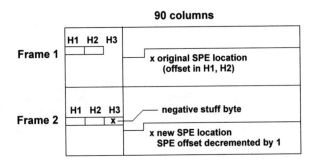

Figure 8.4 Using the H3 byte. When a source clock is fast the H3 byte functions as a negative stuff byte

The A1 and A2 bytes represent framing bytes and denote the beginning of an STS-1 frame. The composition of the bytes is 1111 0110 and 0010 1000. Since they appear horizontally, another common term used for those bytes is horizontal sync pulse.

The J0 byte represents a section-level trace (J0) byte. This byte is used to verify that patching operations are performed correctly.

The B1 byte is a bit interleaved parity code (BIP-8) byte. Even parity is used to check for transmission errors over a regenerator section with its value calculated over all bits of the previous STS-N frame. Thus, the B1 byte provides an error performance monitoring capability for a SONET section.

The E1 byte represents a section-level order wire for voice communications between regenerators, hubs and remote terminal locations. The use of this byte provides a 64 kbps PCM voice channel.

The F1 byte represents a section user data channel byte. This byte provides a 64 kbps user data channel that can be read from and/or written to where each section has terminating equipment.

Bytes D1 through D3 form a section data communications channel that operates at 192 kbps. This channel is used for OAM&P between section terminating equipment.

Line overhead bytes

As indicated in Figure 8.5 there are 18 line overhead bytes associated with an STS-1 frame. Since we previously described the use of the H bytes we will simply note that H1 and H2 are the SPE alignment pointer and H3 is a negative SPE stuff position.

The B2 byte functions in a similar manner to the B1 byte; however, it provides an error performance monitoring capability at the line layer.

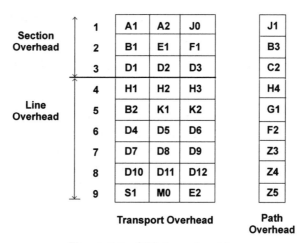

Figure 8.5 STS-1 overhead bytes

The K1 and K2 bytes are referred to as automatic protection switching (APS) bytes. They provide the capability to rearrange a SONET ring as well as to detect alarm indication (AIS) and Remote Detection Indication (RDI) signals.

The D4 through D12 bytes provide a line-level OA&M data channel. Through the use of nine bytes a data rate of 576 kbps is supported.

The S1 byte is a synchronization status byte. Bits 5 through 8 in the S1 byte are used to transport the synchronization status of the network element.

The M0 byte is a line remote error indication (REI). If a receiver receives a corrupt signal it sets the M0 byte to indicate this situation.

The last line overhead byte, E2, represents a 64 kbps PCM order wire byte. This byte provides a voice channel that can be used by technicians but which is ignored as it passes through regenerators.

Path overhead bytes

There are nine bytes in an STS-1 path overhead signal. Those bytes are primarily designed for the payload user, which is normally a communications carrier but which could be an organization that contracts for a private line.

The J1 byte represents an STS path trace byte. A string is repetitively transmitted which enables the receiving terminal to verify it is connected to the transmitting terminal.

The B3 byte contains a path bit interleaved parity code. This byte provides a mechanism for path-level performance monitoring.

The C2 byte is referred to as the STS path signal label byte. This byte indicates the type of payload or content of the STS SPE.

The G1 byte is the path status byte. This byte is used to convey path error reporting back to the original path terminating equipment.

The H4 byte is a multiframe indicator byte. This byte is meaningful for certain payloads.

The Z3 through Z5 bytes which terminate path overhead are presently reserved for future use. Now that we are familiar with the overhead associated with an STS-1 frame, let's turn our attention to the formation of an STS-N signal.

The STS-N frame

An STS-N represents a specific sequence of $N \times 810$ bytes. The resulting frame is formed by byte interleaving of STS-1 frames. Although the transport overhead of nine rows by three columns for each STS-1 frame is aligned prior to interleaving, the associated SPEs do not have to be aligned. This is because each STS-1 includes a pointer to indicate the location of the SPE. Figure 8.6 illustrates the general STS-N frame structure. Note that the SPE floats within the STS-N envelope capacity.

The STS-3 frame

The first level of buildup of the STS-1 signal is obtained by byte interleaving three such signals. The resulting STS-3 signal can be viewed as a frame consisting of nine rows and 270 columns.

Figure 8.7 illustrates the overhead and payload bytes transported within an STS-3 frame. Note that the first three bytes represent the first framing byte (A1) of each of the byte interleaved STS-1 signals, followed by the second framing byte (A2) in a similar manner. Thus, a unique six-byte pattern of three A1 and three A2 bytes marks the beginning of an STS-3 frame. The two Z0 bytes represent an alternative use of the section overhead J0 byte when an STS-N frame is formed. That is, the Z0 byte is placed in the second through Nth position where the J0 byte would normally be located. The use of the Z0 byte is currently allocated for future growth. Similarly, the Z1 and Z2 bytes represent an alternative use of the S1 and M0 bytes in row 9, columns 1 and 2 of the transport overhead area of an STS-1 frame. That is, the Z1 byte is positioned in the second through Nth position of an STS-N frame where the S1

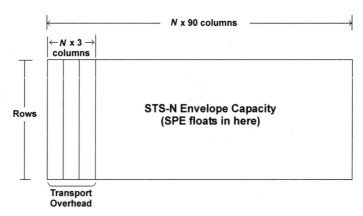

Figure 8.6   The STS-N frame structure

byte would normally fall. The Z2 byte is located in the first and second position where the M0 byte normally goes in an STS-1 of an STS-3, and the first, second, and fourth through Nth positions where the S1 byte normally goes in an STS-1 of an STS-N ($12 \leqslant N \leqslant 48$). Both the Z1 and Z2 bytes are allocated for future growth.

The M1 byte shown in the third STS-1 is used for a Line Remote Error Indication (REI-L) and functions the same as the M0 byte in an STS-1 signal. The payload areas are also byte interleaved. Although each payload byte shown in Figure 8.7 has a distinct location since the payload floats within the payload area, the actual positions of the SPEs are defined by the pointer in the H1 and H2 byte pair associated with each STS-1 signal.

Payloads and VTs

SONET was designed to transport and switch a variety of payloads within an STS-1 signal. Those payloads include a range of DS and European CEPT digital signals that are used to fill an STS-1 signal. Although the non-SONET digital hierarchy involves packing DS1, DS1C and DS2 signals into a DS3 signal, it is often desirable to place such signals in individual locations within an STS-1 signal. Doing so enables individual DS0s to be extracted without requiring an entire DS3 signal to be first decomposed into its individual 28 T1 signals or seven DS2 signals. The actual placement of a North American DS or European CEPT signal into a SONET or SDH frame is accomplished through the use of a virtual tributary (VT), referred to as a virtual container (VC) when working with an SDH STM frame.

	Transport Overhead									Payload				
Section Overhead	A1₁	A1₂	A1₃	A2₁	A2₂	A2₃	J0	Z0	Z0	P1₁	P1₂	P1₃	P2₁	· ·
	B1	–	–	E1	–	–		–	–	P88₁	P88₂	P88₃	P89₁	· ·
	D1	–	–	D2	–	–		–	–	P –	–	·	·	· ·
Line Overhead	H1₁	H1₂	H1₃	H2₁	H2₂	H2₃	H3₁	H3₂	H3₃	P –	·	·	·	· ·
	B2₁	B2₂	B2₃	K1	–	–	K2	–	–	P –	·	·	·	· ·
	D4	–	–	D5	–	–	D6	–	–	P –	·	·	·	· ·
	D7	–	–	D8	–	–	D9	–	–	P –	·	·	·	· ·
	D10			D11			D12			P –	·	·	·	· ·
	S1	Z1	Z1	Z2	Z2	M1	E2	–	–	P –	P –	P –	P –	· ·
	1	2	3	1	2	3	1	2	3	1	2	3	1	

|← 9 columns →|← 261 columns →|

Figure 8.7 The composition of an STS-3 frame

VTs and VCs represent SONET and DSH synchronous formats at sub-STS-1 levels. Both VTs and VCs enable the direct multiplexing of digital signals at rates less than DS3 or CEPT4 into a SONET STS-1 or an SDH STM-0 payload envelope. In actuality there is no such signal as an SDH STM-0. However, as we will note later in this chapter, we can define a Synchronous Transport Module (STM) level 0 signal as the building block for the first defined SDH signal, STM-1. Table 8.2 lists the tributary capacity for different types of VTs and VCs.

VT mappings

SONET supports several methods for mapping a payload into a VT. First, there are three different methods used to support floating mode mapping — asynchronous, bit-synchronous and byte-synchronous.

The asynchronous mode results in a DS1 bit stream not being in synchronization with the VT SPE. This means that bit stuffing must be employed for justification and the DS1 signal is treated as a serial bit stream without any attempt made to identify the byte structure of the signal.

The bit-synchronous mode results in a DS1 signal being synchronized with the VT SPE at the bit level. Although no bit stuffing is required, the DS1 signal is also treated as a bit stream without any attempt being made to identify the byte structure of the signal.

The third method used to support floating mode mappings is the byte-synchronous mode. In this mode a DS1 bit stream is synchronized with the VT SPE at both the bit and frame levels.

Table 8.2 VT/VC tributary capacity.

Signal type	Bit rate (Mbps)	VT/VC type	Possible number in group	Number per STS-1/STM-0 SPE
DS1	1.544	VT1.5	4	28
CEPT 1	2.048	VT2	3	21
DS1C	3.152	VT3	2	14
DS2	6.312	VT6	1	7

This results in the alignment of a DS1 signal with the byte positions of a VT SPE. Only the byte synchronous mode supports the extraction of individual DS0s from a DS1 signal. Each of the previously described methods uses the H1 and H2 byte pointers as well as the H3 stuff byte. Secondly, under a locked-mode option VTs can bypass the use of pointers and obtain a fixed byte mapping into the payload.

VT formation rules

When an STS-1 SPE is used to transport VTs there are a number of rules that must be complied with. First, any given VT group is restricted to transporting one or more tributaries of a single type; however, different VT groups in the same STS-1 can be used for different tributary types. Secondly, each VT group is allocated 12 columns of the SPE, restricting a VT group to one of the following combinations:

- Four VT1.5s (3 columns per VT1.5)
- Three VT2s (4 columns per VT2)
- Two VT3s (6 columns per VT3)
- One VT6 (12 columns per VT6)

Figure 8.8 illustrates the structure and size of each of the four different types of VTs. It should be noted that the 12 columns in a VT group are not placed consecutively within the SPE. Instead, they are interleaved on a column-by-column basis with respect to the other VT groups. In addition, column 1 is used for path overhead while two columns (30 and 59) are not used and are referred to as 'fixed stuff'.

When the SPE of an STS-1 frame is used to transport virtual tributaries, 84 of the 87 payload columns are used. The three unused columns include the path overhead column and two fixed

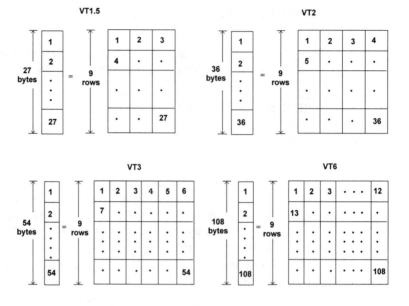

Figure 8.8 The structure and size of virtual tributaries

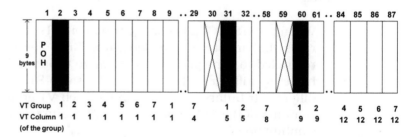

There are seven VT groups, each consisting of 12 columns and nine bytes per column, resulting in 108 bytes

Key: ⊠ stuff byte (not used)

■ used for VT group 1

Figure 8.9 The STS-1 virtual tributary

stuff columns. Figure 8.9 illustrates the format of an STS-1 virtual tributary. In examining Figure 8.9 note that the columns are divided into seven VT groups, with 12 VT columns in each group. Since each column consists of nine bytes per frame, each group provides 108 bytes of transmission capacity. Also note that the shaded columns illustrate how the assignment of a VT1.5 tributary carrying a DS1 signal is accomplished. In this example the use of three columns by nine bytes per column results in the assignment of 27 bytes per STS-1 frame.

VT path overhead

Similar to an SPE, each VT contains a series of path overhead bytes. In the case of a VT there are four evenly distributed path overhead bytes. VT path overhead provides communications between the location where the VT SPE was created and the point where the VT is disassembled. The byte designators for VT path overhead are V5, J2, Z6 and Z7. The first byte (V5) represents the location pointed to by a VT payload pointer and provides the same functions for VT paths that the B3, C2 and G1 bytes provide for STS paths — error checking, signal label and path status. The J2 byte is a VT path trace byte, while Z6 and Z7 represent growth bytes. The J2, Z6 and Z7 bytes occupy the location pointed to by the VT payload pointer in frames 2, 3 and 4 of each sequence of four VT frames. The sequence of four VT frames is commonly referred to as a VT superframe.

OC and EC signals

SONET supports both optical and electrical signals. The standard SONET optical carrier signal (OC-N) utilizes a non-return to zero signal. When optical power is turned on it signifies a binary 1, while a lack of optical power defines a binary 0.

A wavelength of 1310 or 1550 nanometers is specified for SONET optical signaling. The average power of light on a SONET OC should be between -5 and $0\,dBm$. This power is computed by adding the power used to define a binary 1 to the almost-off power level used to define a binary 0 and dividing the sum by two.

There are presently two electrical signals defined for SONET — EC-1 and EC-3. EC-1 uses a Binary 3 Zero Substitution (B3ZS) coding scheme with pulses following the Alternate Mark Inversion (AMI) format. The EC-1 signal is transmitted at $51.84\,Mbps$ via a 75 ohm coaxial cable and is very similar to a DS3 signal.

The EC-3 signal is designed to support transmission at $155.52\,Mbps$. To accomplish this the EC-3 signal uses a Coded Mark Inversion (CMI) line code. Figure 8.10 illustrates an example of the CMI line code. Note that binary 1s are represented by pulses of alternating polarity that have a full 100% bit duration. In comparison under coding pulses are placed in the middle of the bit position. For binary 0 the CMI line code uses a half-period interval

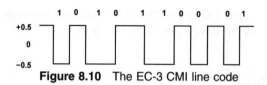

Figure 8.10 The EC-3 CMI line code

of positive voltage. Similar to an EC-1 signal, an EC-3 signal operates via a 75 ohm coaxial cable.

8.3 THE SDH HIERARCHY

As previously discussed, Synchronous Digital Hierarchy (SDH) represents an international standard which is very similar to SONET. SONET begins with an STS-1 signal operating at 51.840 Mbps, while SDH commences with an STM-1 (Synchronous Transpond Module-1) signal which corresponds to the SONET STS-3 signal.

Table 8.3 lists the DSH hierarchy. In examining the entries in that table note that, as previously discussed, there is actually no STM-0 signal designation. However, since an author defined STM-0, which can be considered equivalent to an STS-1 and provides the building block for the construction of an STM-1 signal, it was so designated for illustrative purposes.

The STM-0 frame structure

To achieve a degree of compatibility between SONET and SDH, the STM-1 signal can be considered to represent three STS-1 signals, which results in an operating rate of 3×51.84 Mbps or 155.52 Mbps. As previously noted, the three lower-layer signals used to form an STM-1 signal are not actually defined as a separate SDH signal. However, since those signals represent the basic building block of the STM-1 signal we can refer to those signals as STM-0s. As such, the key difference between an STM-0 and an STS-1 signal is primarily one of terminology.

Table 8.3 The SDH hierarchy.

Signal	Buildup	Bit rate (Mbps)
STM-0	—	51.84
STM-1	$3 \times$ STM-0	155.52
STM-4	$4 \times$ STM-1	622.08
STM-16	$16 \times$ STM-1	2488.32
STM-64	$64 \times$ STM-1	9953.28

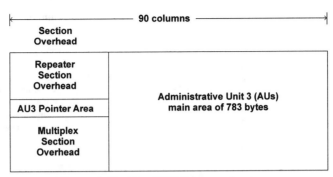

Figure 8.11 The STM-0 frame structure

Figure 8.11 illustrates the STM-0 frame structure. As you will note from a comparison of the STS-1 and STM-0 frame structures, they have near identical structure but employ a different nomenclature. That is, the repeater section is referred to as the section overhead area in an STS-1 frame, while the multiplex section is referred to as the line overhead section in an STS-1 frame. Another difference in nomenclature involves the payload area. In an STM-0 frame the payload area is referred to as administrative unit 3 (AU3). Since the payload floats, the H pointers used in the STS-1 frame were renamed as the AU3 pointer area.

The STM-1 frame structure

Similar to the manner by which three STS-1 signals build an STS-3 frame, three STM-0 signals, each in actuality existing as a virtual signal, can be viewed as forming an STM-1 frame.

Figure 8.12 illustrates the composition of an STM-1 frame. Note that it is similar to an STS-3 frame, the difference being the nomenclature used to describe different components of each frame.

8.4 NETWORK TOPOLOGIES

SONET and SDH support a variety of network topologies, ranging from point-to-point line structure to sophisticated self-healing ring architectures. Non-ring topologies can be used by subscribers to access a communication carrier's ring, or subscribers can elect to have the carrier install a ring directly to their building to obtain a higher level of fault tolerance capability. In this section we will examine four distinct types of network configurations supported by SONET and SDH.

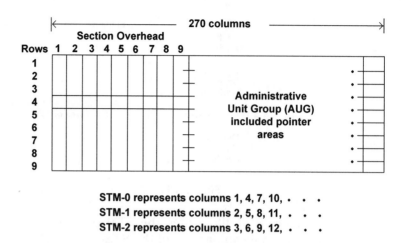

Point-to-point

The most common network topology supported by SONET and SDH is a point-to-point network configuration. The point-to-point network configuration serves as both a building block for constructing rings as well as an access mechanism into a ring.

Figure 8.13 illustrates a point-to-point network configuration. Here the path terminating terminal multiplexer (PTTM) functions as a multiplexer of DS1s, E1s and other tributaries. In its simplest method of deployment a point-to-point configuration includes two PTTMs linked by fiber and may or may not require a regenerator.

Point-to-multidrop

Through the use of an add/drop multiplexer (ADM) you can add and remove circuits as data flows between two locations connected by a point-to-point network configuration. Figure 8.14 illustrates the point-to-multipoint network configuration. Because data flow is via SONET or SDH instead of via a higher-level DS or E signal, this configuration alleviates the necessity to first demultipex, then add and drop channels, followed by remultiplexing of data.

Hub

A hub network configuration resembles a star in that individual point-to-point connections flow to a central location. At that location a Digital Cross Connect System (DCCS) is used to route data between individual SONET and SDH transmission lines.

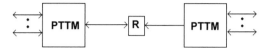

Figure 8.13 A point-to-point network configuration. PTTM stands for path terminating terminal multiplexer

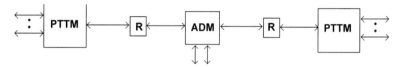

Key: **PTTM Path Terminating Terminal Multiplexer**
     ADM   Add Drop Multiplexer
     R   Repeater

Figure 8.14   Point-to-multipoint network configuration

Figure 8.15 illustrates a hub network configuration. Because all data flows through the DCCS you can monitor network utilization and performance from a single location. In addition, this architecture is well suited to supporting unexpected network growth as well as providing a degree of flexibility for changing a portion of the network configuration.

Ring

One of the design goals of SONET is survivability. To accomplish this goal SONET supports a self-healing feature which enables a fiber cut to be immediately recognized and compensated for by switching traffic onto a different path. That path is normally one of a pair of rings that are constructed as part of a SONET network topology. However, it is important to note that there are two types of SONET rings — one that switches individual paths, referred to as path switched, and one that switches the entire capacity of the optical line and is referred to as line switched. Path-switched rings only use two fibers and are less expensive to construct than a line-switched topology which can utilize either two or four fibers. In addition, as we will soon note, a line-switched ring can provide an additional level of capacity over a path-switched ring.

Path-switched ring

A path-switched ring transmits all traffic both ways around the ring for redundancy. For each path between two nodes the transmitted

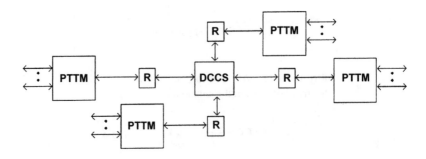

Key: PTTM Path Terminating Terminal Multiplexer
DCCS Digital Cross Connect System
R   Repeater

Figure 8.15 A hub or star network configuration

data flows both clockwise and counter-clockwise, resulting in the use of the term bidirectional ring. The receiving node monitors both signals and selects the better one. Thus, any single break in the transmission medium will enable communications to continue.

Figure 8.16 illustrates a bidirectional ring employing path switching. Due to the dedicated use of alternate paths, path-switched rings have a lessor backup capacity than line-switched rings.

Line-switched ring

A line-switched topology results in the ability to switch at any node in the event of a fiber failure. This capability permits fiber to be inadvertently ripped up between a pair of nodes without bringing down the ring. This is because the detection of a fiber cut can be compensated for by routing traffic around the failure in the opposite direction. Thus, this topology permits the sharing of a SONET or SDH ring structure protection. Figure 8.17 illustrates the construction of a bidirectional line-switched ring topology. Note that one pair of fiber is idle but available for instantaneous use.

The actual switching on a line-switched ring can be accomplished at either the line or the path level. In comparison, path switching on a path-switched ring always occurs at the path level. Another difference between the two types of rings involves the number of fibers used. A line-switched ring supports two and four fibers while a path-switched ring supports two fibers. When a two-fiber line-switched ring is used, only a single pair is used to form

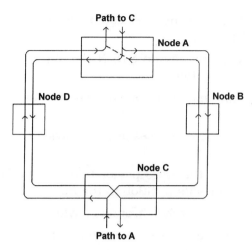

Figure 8.16 A SONET path-switched bidirectional ring

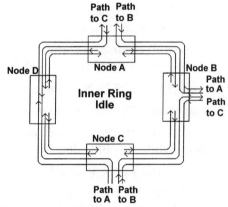

Figure 8.17 A SONET line-switched bidirectional (4-fiber) ring

the ring. However, its capacity is partitioned into two equal portions, with each treated as if it was a separate line. Under this topology line switching occurs on a path basis.

REVIEW QUESTIONS

1 What were the three objectives of the Exchange Carriers Standards Association (ECSA) which were satisfied by the development of SONET?

2 What is the primary difference between SONET and SDH with respect to their operating rates?

3 What are the three components associated with the SONET line structure?

4 Explain why an STS-N signal $(1 \leqslant N \leqslant 192)$ is a multiple N of 51.84 Mbps.

5 How many bytes are used for overhead in a SONET STS-1 frame?

6 What is the gross and net payload, in bytes, of a SONET STS-1 frame?

7 Since the gross payload area in the SONET frame floats, what defines its position in the frame?

8 What is a Synchronous Payload Envelope (SPE)?

9 What defines the position of the SPE within an STS-1 frame?

10 What is the purpose of the H3 byte in the STS-1 transport overhead area?

11 In SONET B bytes are defined for section, line and path overhead. What is the function of those bytes by overhead area?

12 How can technicians communicate with one another if they are located at opposite ends of a SONET section?

13 What does an STS-N frame represent?

14 How is an STS-3 frame formed?

15 Why is the transport overhead section aligned in an STS-N frame while the SPE does not have to be aligned?

16 What is the purpose of Z bytes?

17 What are the three methods used to support floating mode mappings into a VT? Which one supports the extraction of DS0s from a DS1 signal?

18 What is a VT superframe?

19 Over what type of cable would you connect data transmission equipment to an optical transmitter/receiver operating at 155.52 Mbps?

20 Discuss the relationship between an STM-0 frame and an STS-1 frame.

21 Describe four SONET/SDH network architectures. Which one provides the highest level of fault tolerance?

22 What is the difference between a path-switched and a line-switched ring?

9

TESTING AND TROUBLESHOOTING

While digital transmission facilities can be expected to provide a high level of availability and reliability, they can also be expected to be less than problem free. Due to this, the primary objective of this chapter is to focus attention upon the methods by which digital transmission facilities can be tested and how different types of problems can be isolated, a procedure commonly referred to as troubleshooting. To accomplish this objective we will first examine a variety of performance measurements associated with the use of digital facilities. This will be followed by an investigation into the operation and utilization of several types of test equipment as well as the use of equipment indicators for isolation problems. Due to the growth in the utilization of T1 circuits, we will review the electrical specifications and service objectives of that facility that can be used as a basis for testing and troubleshooting. In concluding this chapter we will note SONET and SDH performance measurement metrics as well as appropriate references which denote specific optical measurements.

9.1 PERFORMANCE MEASUREMENTS AND OBJECTIVES

One of the key differences between analog and digital transmission facilities is in the area of performance measurements and objectives. Analog line quality is normally expressed by such terms as signal-to-noise ratio, dB levels, phase (delay) distortion and frequency response. In comparison, the quality of a digital line is normally expressed in terms of the number of errors occurring per unit of time or by the number of units of time in which one or more errors occurred.

Many communications carriers design their digital facilities to provide a level of service in terms of percent error free seconds, number of severely errored seconds, error seconds in an eight-hour day, bit error rate, availability or unavailability percentage. In this section we will first examine the performance measurements associated with digital facilities and then investigate the performance objectives associated with the use of those facilities.

Table 9.1 lists commonly used performance measurements and alarms associated with the use of digital transmission facilities. In the top portion of the table, performance measurements related to a common measurement are so indicated by being indented under a specific measurement.

Performance measurements

Although Table 9.1 lists 16 common performance measurements, in many situations only a small subset of those measurements may be of concern, because of the method by which some communications

Table 9.1 Performance measurements and alarms.

Performance measurements
  Bipolar violation rate
  Bit errors
    Bit error rate
  CRC errors
  Parity errors
  Delay time
  Errored seconds
    Error-free seconds
    Percent error-free seconds
    Unavailable seconds
    Severely errored seconds
    Consecutive severely errored seconds
    Degraded minutes
  Framing errors
    Severely errored frames
  Pattern slips

Alarms
  Loss of clock
  Loss of synchronization
  Yellow alarm
  Alarm indication signal

carriers denote performance on their facilities. By first examining each of the performance measurements listed in Table 9.1, we will obtain a basic understanding of common measurements that will be applicable to most digital facilities.

Bipolar violation rate

A bipolar violation (BPV) occurs whenever two successive 1s have the same polarity. Although the bipolar violation rate can indicate a degree of poor line quality or the presence of defective repeaters, this rate can also be misleading. To understand the latter, consider how several coding formats previously discussed in this book are used to provide a minimum 1s density through the use of intentional bipolar violations. Thus, using test equipment to monitor BPVs, when such equipment counts both intentional and unintentional bipolar violations, can result in the equipment measuring a correct BPV rate that has no relation to line quality nor to the operational status of repeaters. A second problem associated with bipolar violations is the fact that equipment, including multiplexers, CSUs and digital switches, removes bipolar violations. Thus, a BPV rate is only applicable and useful as a measurement for one section of a digital facility.

Under AT&T's Accunet T1.5 service, an excessive bipolar violation condition is considered to occur when any digital circuit experiences a number of bipolar violations that results in a performance level below a threshold of a 10^{-6} BPV rate for 1000 consecutive seconds. This error rate corresponds to 1544 BPVs in 1000 consecutive seconds, since the line rate is 1.544 Mbps.

An excessive bipolar violation condition may indicate a problem with end-user CSUs or a line or repeater problem that should be reported to the carrier for corrective action. To isolate the cause of the problem to end-user or carrier facilities, end-users can consider operating the self-testing feature built into many CSUs. This self-testing feature causes the internal circuitry of the device to be tested and, if successful, indicates that the problem is elsewhere. Thus, upon the successful self-testing of CSUs at both ends of a T-carrier, an excessive bipolar violation rate should be reported to the communications carrier.

Bit errors

A bit error represents the change of a 0 bit to a 1 bit, or of a 1 bit to a 0 bit. By counting bit errors and computing a bit error rate based upon the total number of bits transmitted or the total number of bits transmitted during a predefined period of time, one can obtain a bit error rate. Thus, the bit error rate (BER) can be expressed as

$$\text{BER} = \frac{\text{bits in error}}{\text{total number of bits transmitted}}$$

The BER is used by several communications carriers as a measure for judging circuit quality and availability based upon an ITU recommendation discussed later in this chapter.

In Table 9.2 you will find a list of bit error rates. The top portion of the table indicates the bit error rate equivalents for bit error rates expressed as a power of 10. In the lower portion of Table 9.2, T-carrier bit errors were computed for North American and European circuits operating at 1.544 Mbps and 2.048 Mbps respectively. To illustrate the manner by which the entries in the lower portion of Table 9.2 were performed, consider the error rate of 10^{-9}. At a T1 rate of 1.544 Mbps this is equivalent to 1×10^9 bits$/1.544$ Mbps, or 647.66 s. When divided by 60 seconds per minute we obtain 10.79 minutes or approximately 1 error per 11 minutes.

Table 9.2 Bit error rates.

Bit error rate equivalents

10^{-3}	1 error in 1000 bits
10^{-4}	1 error in 10 000 bits
10^{-5}	1 error in 100 000 bits
10^{-6}	1 error in 1 000 000 bits
10^{-7}	1 error in 10 000 000 bits
10^{-8}	1 error in 100 000 000 bits
10^{-9}	1 error in 1 000 000 000 bits

T-carrier bit errors.

	Bit errors	
Error rate	T1 (1.544 Mbps)	CEPT PCM-30 (2.048 Mbps)
10^{-9}	1 error per 10.79 minutes	1 error per 8.14 minutes
10^{-8}	1 error per 65 seconds	1 error per 48.8 seconds
10^{-7}	1 error per 6.5 seconds	1 error per 4.88 seconds
10^{-6}	1.544 errors per second	2.048 errors per second
10^{-5}	15.44 errors per second	20.48 errors per second
10^{-4}	154.4 errors per second	204.8 errors per second
10^{-3}	1544 errors per second	2048 errors per second

CRC errors

CRC error checking is applicable to extended superframe (ESF) T-carrier and European PCM-30 E-carrier systems using an optional CRC-4 checking algorithm. Under both framing formats, cyclic redundancy checking (CRC) is calculated on the transmitted data and inserted into a portion of the framing information. When the data checked by the CRC algorithm is received, the CRC is recalculated and compared to the original CRC. If the two CRCs match, the transmitted data is considered to have been received without error. If the two CRCs do not match, one or more bits in the frame covered by the CRC check are considered to have been received in error.

A CRC error count provides a mechanism for counting frame errors. Although it does not provide a precise measurement of error activity as bit errors do, CRC errors can be counted without disturbing the flow of data. In comparison, bit error testing requires injecting a known signal onto a line or channel and comparing the received data to the same bit generation process used to generate the bit sequence. This type of testing interrupts the flow of traffic and is known as intrusive testing.

Parity errors

As previously noted in Chapter 5, the C-bit T3 framing format uses parity bit checking as a tool for providing a performance monitoring capability. Similarly, SONET and SDH use B bytes at the line, section and path levels to provide a performance monitoring capability. For T3, SONET and SDH you can monitor the C bits and B bytes without having to take the transmission facility out of service.

Delay time

The routing of data through an extensive digital network can result in cumulative delays. Such delays include the time required to switch channels or packets at nodes, T-carrier multiplexer processing time, and propagation delay time. As delay time increases, its effect upon voice and data transmission become significantly different.

If the delay time exceeds approximately 125 to 250 milliseconds, a voice conversation will start to appear awkward, although a listener will still be able to ascertain what was said. If a data transmission session is occurring, an increasing delay time can become the governing factor resulting in protocol time-outs. In such situations the protocol procedure requires a response to each transmitted block of information within a predefined time period. If this response is not received, the protocol may be configured to drop the session, resulting in a failure to communicate over an operational circuit due to an excessive delay time.

Framing errors

A framing error is a logical error that occurs within the framing bits of a digital signal. Under ESF framing, the occurrences where two or more frame errors transpire within a 3-millisecond multiframe are known as a severely errored frame.

Some test sets can be used to count frame errors. Doing so will provide an indication of a bit error rate per 193 or 256 bits if you assume bit errors are randomly distributed. Of more importance, a framing error count will provide an indication of how frequently T-carrier equipment loses synchronization with one another.

Errored seconds

An errored second (ES) is a second during which one or more bit errors occurred. This measurement forms the basis for several other types of related performance measurements, including error-free seconds, severely errored seconds, consecutive severely errored seconds, unavailable seconds, and degraded minutes.

An error-free second (EFS) is a one-second period of time in which no bit errors occurred. The percentage of error-free seconds then becomes

$$\text{error free seconds (\%)} = 100\% - \frac{\text{error seconds}}{\text{total seconds}} \times 100$$

The percentage of error-free seconds is commonly used as a measurement for qualifying digital circuits at installation as well as providing a basis for the measurement of circuit quality.

A severely errored second (SES) is considered to be any second with a bit error rate greater than or equal to 1×10^{-3}. When monitoring CRC errors where the CRC errors represent a frame or block error rate, a severely errored second then represents any second with 320 or more CRC-6 errors. When two or more severely errored seconds occur consecutively, this situation is known as a consecutively severely errored seconds (CSES) event.

Two additional performance measurements related to errored seconds are unavailable seconds (UA) and degraded minutes (DM). An unavailable second is one second in a period of time in which 10 or more consecutive severely errored seconds occurred. Under ITU recommendation G.821, which is discussed later in this chapter, the occurrence of this event signifies a degree of line deterioration such that a circuit is considered unavailable for use. A degraded minute is a period of 60 consecutive seconds in which the bit error rate exceeded one error per million bits (1×10^{-6}).

Pattern slips

A pattern slip is the addition or deletion of a bit to a transmitted pattern, such as framing. Pattern slips typically indicate a problem with system timing or repeaters on a digital span.

Alarms

In comparison to performance measurements that provide knowledge about the quality level of facilities and equipment, alarms denote the occurrence of abnormal conditions whose cause must be rectified if communications are to be restored.

Types of alarms

A loss of clock occurs when a device has received a long string of consecutive zeros, while a loss of synchronization occurs when two or more of five framing bits are received in error. Both the yellow alarm and alarm indication signal (blue alarm) are indicated by predefined bit patterns previously covered in Chapter 6.

Alarm simulation

To ensure equipment operates correctly, it is often a good idea to use a digital test set as an alarm simulation device. Doing so enables you to observe the response of equipment to simulated alarm conditions and verify if the equipment responds correctly.

9.2 PERFORMANCE CLASSIFICATIONS

The ITU G.821 recommendation defines four error rate performance categories. These four categories are available and acceptable, available but degraded, available but unacceptable, and unavailable, and are graphically illustrated in Figure 9.1

Availability levels

The available and acceptable performance category is based upon intervals of test time of at least one minute during which the bit error rate is under 10^{-6}. This performance category results in a good voice quality on a T-carrier facility.

The available but degraded performance category is based upon intervals of test time of at least one minute during which the error rate is between 10^{-3} and 10^{-6}. The occurrence of this performance category usually results from microwave fading, atmospheric disturbances or degraded repeaters.

The available but unacceptable performance category is based upon intervals of test time of at least one second, but less than 10 consecutive seconds, during which the error rate is greater than 10^{-3}. On a T-carrier facility, an unacceptable performance level normally results from error bursts, clock slips and circuit switching hits.

The last performance category defined by the ITU G.821 recommendation is unavailable. This performance category is

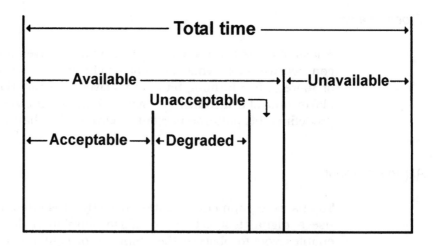

Figure 9.1 Performance classifications

based upon intervals of test time of at least 10 consecutive seconds during which the error rate is greater than 10^{-3}. When this situation occurs, a T-carrier facility is considered inoperative and not available for use.

Computing availability

Over a period of time, circuit availability can be computed and compared to a carrier's guaranteed level of performance, which is normally expressed as a percentage of availability. Since available circuit time is total time less unavailable time, availability expressed as a percentage becomes

$$\text{availability } (\%) = \frac{\text{availability}}{\text{available} + \text{unavailable}} \times 100$$

where

$$\text{available} + \text{unavailable} = \text{total time}$$

Although a high level of availability is very desirable, by itself the figure can be misleading. As an example of this, consider Table 9.3 which lists outages you can expect based upon five levels of availability. Note that total time in computing availability is based upon a 24-hour day, even though most organizations may use a digital facility for only a fraction of that time. Since some T-carrier facilities are still routed over microwave systems, you can normally expect a higher level of electromagnetic interference to occur during daytime. This, in turn, will more than likely result in a lower level of availability during daytime than in other periods of the day.

Currently, AT&T specifies a 99.9% level of availability for its subrate Dataphone Digital Service facilities. For that vendor's Accunet T1.5 1.544 Mbps transmission facilities, a 99.7% level of availability is specified.

Table 9.3 Availability versus outages.

Percent availability	Monthly outage	Annual outage
99.5	3.7 hours	44 hours
99.8	1.5 hours	17.5 hours
99.9	44 minutes	8.8 hours
99.99	4.4 minutes	53 minutes
99.999	27 seconds	5.3 minutes

9.3 COMMON TESTS AND TEST EQUIPMENT

In this section we will examine several common tests and the use of test equipment to obtain an indication of the quality of a digital facility as well as the operation of equipment connected to such facilities.

BERT

Bit error rate testing (BERT) involves generating a known data sequence into a transmission device and examining the received sequence at the same device or at a remote device for errors.

Normally, BERT testing capability is built into another device, such as a protocol analyzer or multiplexer. The use of a BERT results in the computation of a bit error rate (BER), which is

$$\text{BER} = \frac{\text{bits received in error}}{\text{number of bits transmitted}}$$

To perform a bit error rate test using one tester, communications equipment must be placed into a local loop-back mode of operation and the BERT test can then be used to determine if equipment is operating correctly. If a BERT test is conducted end-to-end via a loop-back at the distant end of the circuit, the test can indicate the level of performance of equipment connected to the facility and the facility. In certain situations equipment at the distant end of a circuit can be placed into a loop-back mode of operation in which the digital facilities transmit and receive wires are bridged. When this occurs and a bit error rate tester is directly connected to a DSU or CSU, the resulting measurement can be used as an indication of circuit quality.

Figure 9.2 illustrates how the distant end of a T-carrier circuit can be placed into a loop-back mode of operation. In this example the loop-back can occur at the network interface (NI) installed at the carrier, at the channel service unit (CSU), or at the data terminal equipment (DTE). Loop-backs at the NI or CSU can be effected by sending for a period of five seconds a repeating pattern as denoted in Table 9.4 for North American T1 facilities. To loop-back the DTE, such as a multiplexer, requires either manual intervention at the remote site or if the DTE is operating under the control of a central facility the transmission of an appropriate instruction to place the device into loopback.

Once a loop-back is effected you can use one BERT to perform a BER which can be used to obtain an indication of end-to-end

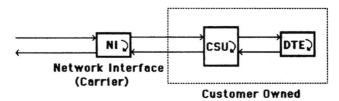

Figure 9.2 Loop-back methods. Remote loopbacks can be effected at the network interface, CSU, or at the DTE

Table 9.4 T1 loop-back codes.

Device	Operation	Bit pattern	Result
CSU	SET	10 000	Places CSU in loop-back
	RESET	100	Cancels CSU loop-back
NI	SET	11 000	Places NI in loop-back
	RESET	11 100	Cancels NI loop-back

channel performance. As an example, a bit error rate in excess of one per 1000 during 10 consecutive seconds would indicate an unavailable level of performance based upon the ITU G.821 recommendation.

The most popular test signal for T1 troubleshooting is a repeating pattern of $2^{20} - 1$ (1 048 575) bits as specified in AT&T publication 62411. Known as a quasi-random signal source (QRSS), it violates minimum density requirements since bits 2 through 49 of the pattern contain 44 zeros and four ones, while a 48-bit string requires at least five zeros.

When used for testing a DS1 signal the QRSS pattern is used to simulate the activity on individual or all DS0 channels. All bits in each frame with the exception of the framing bit result from the generation of a repeating 1 048 575 bit pattern. Some test sets have an option which forces an output bit to a binary 1 whenever the next 14 bits in the pattern will be binary 0s. This option is unnecessary when B8ZS coding is used, since zero suppression is then generated by the CSU that complies with the coding format. Since QRSS testing is disruptive to data traffic, CRC analysis and counting of framing bit errors should be considered. Both of these methods provide a non-disruptive method of obtaining performance data.

Table 9.5 indicates the test time required to calculate bit error rates when a QRSS test pattern is inserted into a single DS0 channel on a T1 circuit. Since the time is proportionally reduced as

Table 9.5 Error rate calculation time per DS0.

Error rate	Calculation time
10^{-3}	1.56 seconds
10^{-4}	15.6 seconds
10^{-5}	156 seconds
10^{-6}	1562 seconds
10^{-7}	28.6 minutes
10^{-8}	51.7 minutes
10^{-9}	286 minutes

the number of DS0 channels carrying the signal is increased, you can divide the calculation time by the number of channels used to transmit the QRSS sequence if more than one DS0 channel is used.

T-carrier BERT plug-in modules

Several T-carrier BERTs use plug-in modules to support different types of T-carrier facilities. Interface plug-in modules govern the physical interface such as RS232, ITU V.35 or RS-449; the method of framing, including unframed, D4, ESF or CEPT PCM-30; the method of line code support, such as B7, B8ZS or HDB3; and countable events, including bipolar violations, framing errors, CRC errors and bit errors.

In its test mode of operation, the BERT simultaneously counts transmitted and received characters and bits in error, and may count bipolar violations, framing errors and CRC errors. If the error type being counted is CRC errors the bit error rate (BER) may represent a block error ratio depending upon the BERT test device used. As an example, some BERTs count the number of errored CRC blocks by considering an extended superframe of 4632 bits to be one block of data. If a bit error occurs within the six CRC bits, the extended superframe is considered to be a block error. Then, the number of errored CRC blocks is divided by the total number of CRC blocks to obtain a block error ratio.

Another interesting aspect of some BERT test sets is the ability to select patterns that generate a violation of 1s density requirements. Then the injection of this type of test sequence can be used to verify the correct operation of equipment performing B8ZS or HDB3 line coding, in effect performing a stress test on that equipment.

To convert an error counter number in a BERT to a bit error rate, testing for a fixed period of time is required. Table 9.6 lists the time

Table 9.6 Bit error rate versus test times.

Data rate (bps)	Bit error rate		
	1×10^{-5}	1×10^{-6}	1×10^{-9}
300	5 min 33 s	55 min 33 s	55 555 min
600	2 min 47 s	27 min 47 s	27 777 min
1200	1 min 23 s	13 min 53 s	13 888 min
2400	42 s	6 min 57 s	6944 min
4800	21 s	3 min 28 s	3472 min
9600	11 s	1 min 44 s	1736 min
19.2k	6 s	52 s	868 min
56k		17.8 s	297 min
64k		15.6 s	260 min
1.544M			10.8 min
2.048M			8.14 min

required for two common bit error rates based upon seven distinct data rates. Note that this table can be used for testing low-speed multiplexer channels as well as currently available subrate digital services.

As an example of the use of Table 9.6, consider the 9600 bps data rate for testing purposes. If, during a test time of 11 seconds, exactly three bit errors occurred, then the bit error rate is 3×10^{-5}. To illustrate the conversion of one bit error rate to another bit error rate, assume a BERT was performed on a PCM-30 system operating at 2.048 Mbps for precisely 8.14 minutes and the bit error counter displays 1631 bits in error. This would indicate a BER of 1631×10^{-9}, or 1.631×10^{-6}.

Since most protocols group data into blocks for transmissions, a block error rate may provide a more realistic level of performance than a bit error rate. This is because a burst of noise resulting in a high bit error rate may affect fewer blocks and result in a lower block error rate than a lower bit error rate where errors are more evenly distributed over time.

Error rate testing methods

Figure 6.3 illustrates three common methods whereby bit error rate testing can be used on digital facilities. In each example the unlabeled rectangles represent either a combined DSU/CSU or a CSU, with the type of device dependent upon whether subrate or T-carrier facilities are being tested.

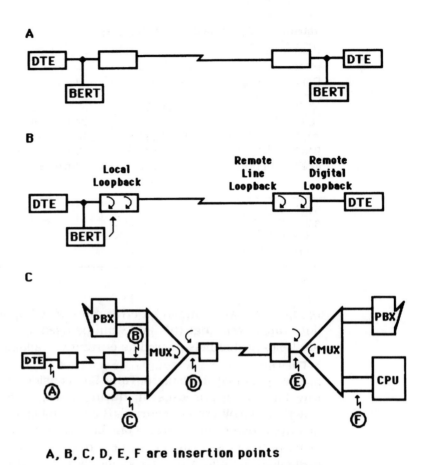

A, B, C, D, E, F are insertion points

Figure 9.3 BERT testing: A end-to-end; B fault isolation using loop-backs; C network component testing

The end-to-end testing illustrated in Figure 9.3A that is performed without a loop-back requires the use of two BERTs. The advantage in using this method of testing is that problems can be isolated to either side of the transmission line.

In Figure 9.3B, the process of performing fault isolation using four types of loop-backs is illustrated. In comparing parts A and B of Figure 9.3, note that through the use of loop-backs you can perform testing with one test set instead of two. In addition, through the use of loop-backs you obtain the ability to test the operational status of both selected equipment and portions of the circuit. To illustrate this let us examine the four loop-backs illustrated in Figure 9.3B.

The first loop-back moving from left to right is commonly called a local digital loop-back. This loop-back ties transmitter to receiver without data being modulated into a bipolar signal and then demodulated back into a unipolar signal. Using this loop-back, in effect, echoes data back to the DTE or test instrument. If a DTE was used by itself when a local digital loop-back was in effect, transmitted characters would be echoed back to the device. Thus, this loopback can be used to test both the operation of the DTE as well as the cable connecting the DTE to the DSU or CSU. Similarly, using a BERT would test the cable connection to the DSU or CSU.

The second loop-back is commonly referred to as a local analog loop even though it results in the DSU/CSU or CSU converting unipolar digital signals to bipolar digital signals and back to unipolar. The reason for the term local analog loop-back is that it refers to the function modems perform when placed in that type of loop-back. Although DSU/CSU or CSUs modulate digital data, the term continues to be used as a hold-over from analog terminology. When placed in a local analog loop-back mode of operation, the BERT tests the complete DSU/CSU or CSU, including its modulation and demodulation circuitry. Note that neither the local digital nor the local analog loopbacks permit the high-speed circuit to be examined.

The third loop-back, commonly referred to as a remote line loop-back or remote digital loop-back, connects the transmit and receive wires of the circuit. This functions as a digital bridge, passing bipolar violations back to the local site without correction. Thus, a test instrument inserted at the local site on the receive side of the line would obtain an accurate count of bipolar violations on the loop when the remote DSU/CSU or CSU is placed into a line loop-back mode of operation.

The fourth loop-back is similar to the second loop-back with respect to the fact that digital data in the line's bipolar format is converted to unipolar data and then back to bipolar for transmission back to the local site. This loop-back loops in the digital section of the DSU/CSU or CSU and converts bipolar violations to binary logic errors prior to encoding the resulting unipolar signal back to a bipolar signal for transmission.

Although the remote line loop-back provides a mechanism to measure the BPV rate, it also provides a mechanism to obtain the BER on the line, since a simple bridge operation is performed by the remote device to tie the circuit's transmit and receive wires together. When a remote digital loop-back is effected, a test from the local site measures not only line quality but, in addition, the operational status of the remote DSU/CSU or CSU.

Both local analog and local digital loop-backs are normally set by the use of switches or buttons on the local device. To effect remote loop-backs, some devices, including locally installed DSU/CSUs and CSUs, will generate appropriate loop codes when switches of buttons labeled 'Remote' are toggled or pressed. Typically, these buttons or switches have two position settings labeled set and reset, with placement to the set position causing an activate loop-back code to be transmitted to the remote device.

Insertion points

When communications equipment, including multiplexers, is connected via digital facilities, both the number of insertion points that can be used for testing, as well as the types of loop-backs one can use for testing, can substantially increase. This is illustrated in Figure 6.3C. Here the curved lines inside each multiplexer indicate the loop-back of a specific channel or channels for testing of lines connected to each multiplexer. The curved lines outside each multiplexer indicate a remote loop-back which can be performed on a channel basis or on the entire carrier circuit. Finally, the circled letters represent a few of the possible insertion points where test equipment can be used in conjunction with the loop-back of a specific multiplexer channel to ascertain the performance of equipment or facilities similar to the methods described previously.

BLERT

Block error rate testing (BLERT) provides a more realistic measurement of data transmission performance than bit error rate testing. This results from the fact that data transmission involves the blocking of data, where one bit in error results in the necessity to retransmit the entire block. Thus, a high bit error rate resulting from a burst of errors might adversely affect one or a few data blocks, while a low bit error rate in which single bit errors occur periodically can adversely affect a large number of blocks.

The block error rate (BLER) is defined as follows:

$$\text{BLER} = \frac{\text{blocks in error}}{\text{total blocks}}$$

Block error rate testing is commonly used on synchronous channels on a T- or E-carrier multiplexer. Many T- and E-carrier multiplexers include both a built-in bit and block error rate testing capability.

Error-free second tester

In an error-free second (EFS) test, received data is analyzed on a per second basis. If one or more bit errors occurs during a one-second interval, the interval is recorded as an errored second.

An error-free second tester can be employed in a manner similar to that described for bit error rate testing. That is, the tester can be used with four types of loop-backs described for DSU/CSU or CSU operations or the tester can be employed using loop-backs that are available through the use of different types of DTEs.

EFS testers incorporate a variety of interface connectors that can include V.35 (34 pin), RS-449 (37 pin), RS-232 (25 pin), or X.21 (15 pin) interface connectors for use on digital facilities. Typical data rates supported by EFS testers include subrates from 2.4 kbps to 56 or 64 kbps, as well as T- and E-carrier 1.544 and 2.048 Mbps operating rates.

As previously discussed in this chapter, many communications carriers specify the availability of their high-speed digital facilities in terms of errored seconds or error-free seconds over a specified time period. Table 9.7 compares the ITU G.821 error performance recommendation with British Telecom's 64 kbps digital service goal, and AT&T DDS and Accunet T1.5 service objectives with the latter for circuits with distances exceeding 1000 miles (1600 km). In actuality, the performance objectives of AT&T's Accunet T1.5 circuits are complex and are based upon both the type of line segment as well as the length of the segment. For customer premises-to-customer premises distances less than 250 miles (400 km), the error-free seconds performance goal is 96.6%, while distances greater than 1000 miles have a goal of 95.0% error-free seconds. For serving office-to-serving office circuit segments the performance objectives are 99.1%, 98.5%, and 97.5% error-free seconds for distances less than 250 miles, distances from 250 to 1000 miles, and distances greater than 1000 miles respectively. For local exchange carrier provided access from the customer premises to the serving office the performance objective is 98.75% error-free seconds.

T-carrier simulator

A T-carrier simulator permits users to predict how a system will perform. The simulator permits users to corrupt the T-carrier framing pattern to determine how a device responds to framing errors, framing errors at a rate less than that needed to cause a loss

Table 9.7 Digital circuit error performance.

	% Error-free seconds	Errors in 8-hour day
ITU Recommendation G.821 for 64 kbps service	98.8	346 s
British Telecom 64 kbps KiloStream/MegaStream goal	99.5	144 s
Dataphone Digital Service	99.5	144 s
Accunet T1.5	95.0	1440s

of frame synchronization, or framing errors at a rate equal or above that necessary to cause a loss of synchronization.

Through the use of a T-carrier simulator, you can determine if yellow and red alarms are generated and if a device under test responds correctly, as well as determine the time required to recover from a loss of synchronization. Other practical uses of a T-carrier simulator include modeling data link impairments under controlled programmable conditions, testing the development of hardware and software off-line prior to their actual installation and operation, and the periodic performance testing of hardware and software.

T-carrier transmission test set

A T-carrier transmission test set is the most sophisticated of all types of test equipment that can be used with digital facilities. Equipment in this category can be used to perform frame level tests, gather statistics concerning different types of error conditions, and may include the ability to remove a channel from a T-carrier line and apply traditional analog or digital testing to that channel. Table 9.8 lists some of the typical measurements T-carrier transmission test sets are capable of performing. Since most of these performance measurements were previously discussed in this chapter, let us focus our attention upon those measurements not previously discussed as well as on certain elements of specific measurement that warrant an elaboration.

Jitter and wander

Both jitter and wander can be considered as a potential disaster waiting to occur. To understand the basis for each impairment it is

Table 9.8 T-carrier transmission test set measurements.

Frame errors and alarms
   Frame bit error ratio
   Severely errored seconds
   Out-of-frame seconds
   Loss of frame (red alarm)
   Loss of frame seconds
   Yellow alarm seconds
CRC-6 errors (ESF) and CRC-4 errors (CEPT PCM-30)
   Average and current BER
   Errored seconds
   % error-free seconds
Jitter and wander
   Jitter hit seconds
   Phase hit seconds
Line and signal measurements
   BPV count
   BPV errored seconds
   Excess zeros
Simulation
   All ones (blue alarm)
   Red alarm
   Yellow alarm

necessary to first understand how timing delays occur in a digital network.

As data pulses flow through a digital network, repeaters recover pulse clocking from the incoming data. Unfortunately, this recovery is not instantaneous as there are built-in delays in the circuitry which recognize an incoming signal and then regenerate the signal. Another source of timing delays is digital multiplexers that build subrate facilities onto a T-carrier circuit. Such multiplexers add bits, a process called bit stuffing, to synchronize the low-speed incoming digital pulses to the T-carrier's operating rate. Since bit insertion will not occur at a precise time due to circuitry delays, the multiplexer is another source of timing delays.

The difference between the ideal and the actual time of arrival of a digital pulse is known as jitter and wander, the term used depending upon the magnitude of the difference. If the difference is at a rate less than 10 Hz it is known as wander, while a difference between the ideal and actual arrival time of a pulse that exceeds 10 Hz is known as jitter. Since both wander and jitter are cumulative they will eventually build up to a point where network synchronization will be lost, resulting in a condition in which a bit time is eventually either gained or lost as illustrated in Figure 9.4.

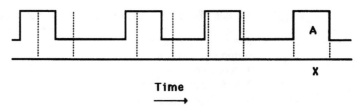

Time
⟶

Figure 9.4 Displacement in time of a signal. Although bit A at time X is in its exact clock position, the cumulative jitter will result in succeeding bits being displaced more and more by time

To compensate for the effect of wander and jitter, digital network elements contain buffers, functioning as a compensation unit between the clocking differences building up in the network. As an example, data entered into a multiplexer at one rate is first buffered prior to transmission at a different rate. Thus, buffers in effect can remove accumulated jitter as long as the variations do not exceed their capacity. When an excessive amount of jitter or wander occurs, the buffers in network equipment may either overflow or underflow. When an overflow condition occurs due to data pulses arriving early, the buffer must delete a block of data to maintain synchronization. In comparison, when data pulses arrive late, this condition results in an underflow in which the buffer must repeat a block of data to maintain synchronization with the T-carrier operating rate. Both of these conditions are known as a slip which can be defined as the occurrence of a digital signal buffer overflow or underflow.

When transmission occurs on T-carrier facilities, digital network equipment will intentionally delete or repeat all or a portion of the bits in a frame. When the network equipment performs a slip by deleting or repeating all frame bits, this process is known as a controlled slip. When the network equipment either deletes or repeats a portion of a T-carrier frame, the process is called an uncontrolled slip.

Controlled slips are performed by T-carrier equipment that performs switching and cross connection functions, such as DACS, PBXs and central office switching systems. Uncontrolled slips result from the underflow or overflow of buffers that are smaller than the size of a frame, hence they are referred to as unframed buffer slips.

Unframed buffers are incorporated into higher rate multiplexers that are not synchronized to a common frequency source, which, in effect, multiplex data sources asynchronously. Examples of equipment with unframed buffers include AT&T M13 and M23 multiplexers as well as the dejitterizing circuitry in end-user T-carrier multiplexers. Unfortunately, an uncontrolled slip can be much more serious than a controlled slip as it results in the shift of the framing bit positions. This shift of the framing bit position is known as a change of frame alignment (COFA) which will result in an out-of-frame condition at the receiving T-carrier multiplexer, causing the multiplexers to resynchronize for a period of time in which data cannot be passed.

ITU recommendation G.822 specifies service objectives for international 64 kbps digital facilities. Under this recommendation there should be five or fewer slips during a 24-hour period for 98.9% of the time, more than five slips per 24-hour period but no more than 30 slips in any hour for less than 1% of the time, and more than 30 slips per hour for less than 0.1% of the time.

Line and signal measurements

To effectively monitor bipolar violations the test set must not count intentional violations as errors, such as B8ZS or HDB3 coding. In addition, care should be taken as to the placement of the test set since CSUs are designed to remove bipolar violations.

Figure 9.5 illustrates the correct insertion of a test set into a digital facility to count BPVs and BPV errored seconds. Note that the test set must have a built-in CSU to transmit bipolar data onto the network at the network interface as well as the ability to generate a remote CSU loop-back code to place the remote CSU into loop-back.

By measuring BPV data you can determine the performance of digital repeaters, as well as obtain information concerning noise and crosstalk on the facility, since those impairments also contribute toward generating BPVs. Since digital radio, satellites and fiber optics do not use a bipolar transmission format, this test, in many instances, indicates premises-to-central office performance and is not a measure of end-to-end performance.

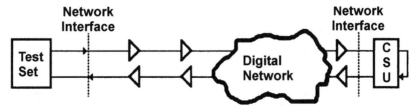

Figure 9.5 Troubleshooting problems at the network interface

9.4 T1 ELECTRICAL SPECIFICATIONS AND SERVICE OBJECTIVES

One of the key problems associated with the testing of a T1 circuit is in attempting to determine if the measured and observed circuit characteristics indicate a problem. In this section we will review T1 electrical specifications and service objectives that can be used by readers as guidelines for testing T1 circuits.

Electrical specifications

In general, the electrical specifications of a T1 circuit define the pulse characteristics of the signal, including its pulse type, pulse rate, pulse shape and pulse density. Concerning pulse type, a T1 circuit uses bipolar return to zero signaling with 0 volts as the base reference. The pulse rate is 1.544 Mbps; however, many carriers permit a deviation of ± 75 bps. Similarly, the 3 volt dc pulse height or amplitude can vary by $\pm 10\%$, while the pulse width of 324 nanoseconds can normally vary by ± 45 ns. Lastly, as previously noted, T1 terminal equipment cannot transmit more than 15 consecutive zeros when B7 zero code suppression is used or a bipolar violation must occur when a byte is all zero when B8ZS coding is supported. Table 9.9 summarizes the electrical specifications of a T1 carrier that can be used for testing guidelines.

Service objectives

The service objectives of a T1 facility are expressed by carriers in a variety of ways, including percent error-free seconds, bit error rate and availability. In general, most carriers have a goal to provide T1 users with a 99% or above level of availability per month and an error-free second (EFS) rate at or above 95% over each 24-hour

Table 9.9 T1 electrical specifications.

Specification	Expected value
Pulse type	Bipolar return to zero
Pulse rate	1.544 Mbps ± 75 bps
Pulse amplitude	3 V dc $\pm 10\%$
Pulse width	324 ± 45 ns
Pulse density	<15 consecutive zeros B7 zero code suppression
	1 bipolar violation per zero byte B8Zs coding

period. While most carriers focus upon EFS and availability, a bit error rate exceeding 1 in 1 000 000 is normally considered poor for this type of facility and deserves reporting even if the carrier does not express service objectives in terms of a bit error rate.

By examining the electrical specifications and service objectives of your facility and comparing observed values to those previously discussed in this section, you may be able to quickly isolate problems. If testing does not indicate any deviations from the previously indicated values, you may then want to consider performing other tests and parameter measurements previously described in this chapter and compare those results against the detailed specifications provided by the communications carrier.

9.5 SONET AND SDH

Both SONET and SDH frames transport performance bytes which enable the quality of the transmission facility to be determined without having to remove the facility from service. However, when a SONET or SDH signal becomes impaired it is no different from a T- or E-carrier signal with respect to the need to perform a number of tests to attempt to isolate the cause of the impairment.

There are two major categories of tests you can consider performing on SONET and SDH transmission facilities — signal and signal structure. A signal test examines the phase of the signal to determine if there is an excessive amount of jitter or wander. In an optical network, just like their electrical cousins, variations above 10 Hz are referred to as jitter while variations below 10 Hz are referred to as wander.

Jitter sources

Common sources of jitter in an optical network include add/drop multiplexers and noise and intersymbol interference. As a signal passes through a repeater it can experience a degree of jitter transfer gain, resulting in an increase in line jitter. As jitter builds up it can reach a level where network equipment cannot tolerate its buildup. Recognizing the fact that jitter generation by itself can be within tolerance but a jitter problem can still exist, ANSI and Bellcore defined specified limits on four jitter parameters. Three of those parameters are illustrated in Figure 9.6 and include jitter generation, jitter transfer and jitter tolerance. The fourth jitter parameter, interface jitter, refers to a system specification.

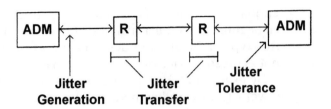

Figure 9.6 Jitter accumulation on an optical network

Jitter generation limits are defined by Bellcore based upon the optical carrier (OC) level. Similarly, jitter tolerance and jitter transfer specifications, as well as interface jitter limits, are defined based upon the OC level. Readers are referred to ANSI T1.105.03-1994 and Bellcore GR-253-CORE publications for detailed information concerning jitter limits.

Wander sources

Wander represents a very slow drift of phase. In a synchronous system wander can result from DC offset drift in phase-locked loops that time one reference from another, frequency differences during a resynchronization process, and even changes in propagation delay in cable due to temperature changes. One of the most common causes of wander results from a T- or E-carrier signal improperly synchronized with an OC signal. Since wander represents low frequencies with long periods, its buildup can occur over a long period of time. To provide a concise measure of wander Bellcore defined three specific measurements referred to as Time Interval Error (TIE), Maximum Time Interval Error (MTIE) and Time Deviation (TDEV).

TIE users a stratum 1 clock as a wander-free reference for measuring the wander of the optical signal in nanoseconds. MTIE provides a measure of wander which defines frequency offsets and phase transients of the signal. The third component of wander measurements, TDEV, characterizes wander with respect to its spectral content. Readers are referred to the previously mentioned Bellcore and ANSI publications for information concerning wander specifications.

Signal structure

A second area of SONET and SDH testing involves verifying their signal structure. Since both standards define multiple signal rates and payload types, a number of tests were developed to verify their signal structure. Three commonly used tests include path verification, line verification and section verification. Each of those tests is accomplished by monitoring the appropriate J byte path trace message discussed in Chapter 8. Another common test is referred to as a pointer sequence test. ANSI, Bellcore and the ITU define specific pointer sequences for STS/STM pointers and VT/VC pointers. Such standards define permissible variances in pointers due to phase transients resulting from synchronization changes. Several test equipment vendors currently provide SONET and SDH test sets that compare current pointer positions to the specific pointer sequences to determine the quality of the signal structure.

REVIEW QUESTIONS

1 How could an excessive bipolar violation rate be a false indication of the quality of a digital circuit?

2 What types of problems could an excessive bipolar violation rate indicate? How could you isolate the problem area to end-user equipment or communications carrier facilities?

3 If the expected error rate on a CEPT-30 facility is 10^{-7}, how many errors can you expect to occur during a one-minute period?

4 Discuss an advantage and disadvantage of CRC errors with respect to bit errors.

5 What is intrusive testing?

6 Describe the effect of an increasing delay time upon voice and data transmission.

7 What is an error-free second?

8 What is the relationship between error-free seconds, error seconds, and total seconds?

9 What is a severely errored second? What is the relationship between a severely errored second and consecutive severely errored seconds?

10 What is an unavailable second?

11 What is a pattern slip and what does its occurrence most likely indicate?

12 Why would you consider using a test set to simulate an alarm condition to a T- or E-carrier multiplexer?

13 Explain why a high level of availability cited by a communications carrier may not be experienced by an organization whose transmission requirements primarily occur between 6 am and 5 pm.

14 What are the three types of loop-backs that can occur on a T1 circuit and how is each loop-back effected?

15 Assume you conducted a bit error rate test on a 128 kbps portion of a T1 circuit. If 88 bit errors occurred during a 51.7 minute period, what is the bit error rate per DS0 channel?

16 Why would you perform a local digital loop-back prior to conducting a BERT test?

17 Why is a BLERT test a better indication of potential data transmission performance than a BERT test?

18 When should you consider using a T1 simulator?

19 What are two sources of timing delays on digital networks?

20 What are the two major types of tests you can perform on SONET and SDH?

10

DIGITAL CIRCUIT
RESTORAL OPERATIONS

Because organizations rely heavily upon high capacity digital transmission circuits, the duration of an outage has a significant effect upon their ability to conduct business. Recognizing this fact, both communications carriers and third party vendors offer several circuit restoral solutions you can consider. These solutions are the focus of this chapter.

In this chapter we will first turn our attention to several communications carrier restoral options you can consider. Once this is accomplished, we will turn our attention to third party vendor and user implemented solutions that can provide a viable backup and recovery mechanism for digital transmission facilities.

10.1 COMMUNICATIONS CARRIER SOLUTIONS

Communications carriers offer a variety of digital circuit restoral solutions you can consider. These range in scope from redundant circuits, route diversity and route avoidance techniques to specific carrier offerings.

Redundant circuits

The installation of redundant circuits can give a false sense of security, because there is a high level of probability that such circuits or portions of them will be routed through the same central office. This means that a central office fire, flood or human error could render both circuits inoperative when you most need them. Thus, it is highly possible that after years of paying double for an additional circuit, it will be inoperative at the moment when you

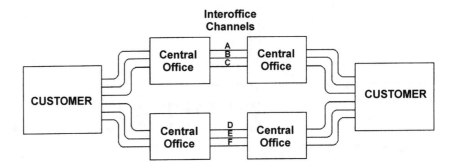

Figure 10.1 Redundant route diversity circuits

most need its transmission capability. A far better method for circuit restoral is the use of redundant route diversity circuits.

Redundant route diversity circuits

Redundant route diversity circuits are a pair of circuits that are entirely distributed over physically separate routes. This route separation begins at the customer's premises, with access lines routed to different central offices. From each central office each circuit is placed on a different interexchange carrier transmission facility for routing to the destination city. At the destination city two central offices route access lines to the customer via different paths, providing two end-to-end separate transmission paths.

Figure 10.1 illustrates an example of redundant route diversity for three pairs of circuits. In this example A–D, B–E and C–F are diversity routed. The cost of diversity routing varies by carrier and type of digital service. Typically the cost of diversity routing includes a one-time fee for installation of separately routed digital circuits and a monthly recurring fee.

Route avoidance

A third type of circuit routing structure offered by some communications carriers is route avoidance. Under route avoidance the customer specifies or requests the communications carrier to avoid a particular geographic area; however, such routing does not provide for the restoration of a failed transmission facility.

The primary reason for selecting route avoidance is to ensure a circuit does not flow through an area experiencing a specific type of problem such as flooding in a specific area of a country.

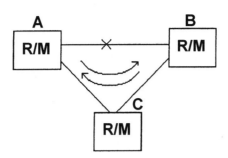

Figure 10.2 Using a router or multiplexer based network to compensate for a circuit failure; R/M = router or multiplexer

Automatic protection capability

Automatic protection capability (APC) represents an AT&T service which provides protection against a T1 or T45 local access channel failure. This protection requires subscribers to install a separate local access channel to serve as a spare as well as automatic switching equipment at their site which is compatible with AT&T switching equipment. Under APC the failure of a local access channel is compensated for by the rerouting of data to a spare channel. In addition to AT&T other interchange carriers offer a similar service under different names.

Network protection capability

Network Protection Capability (NPC) is an AT&T offering which protects the interoffice carrier (IOC) portion of a T1 or T3 carrier through the use of a switching arrangement that automatically switches the subscriber's IOC to a separately routed fiber optic cable in the event that the primary IOC access fails. NPC is similar to diversity routing on the local access portion of a circuit. Similar to APC, several carriers in addition to AT&T offer a NPC type of service.

Reserved service

Under AT&T's Reserved Service a dedicated T1 circuit is brought online after a user orders it with a telephone call. Reserved Service is available by pre-subscription and requires local access facilities to be in place for the service to work.

Customer controlled reconfiguration

Through the use of an on-premises terminal a subscriber can reprogram an AT&T Digital Access and Cross Connect System

Table 10.1 Third party/subscriber circuit restoral solutions.

VSAT transmission
Switched Digital Network usage
T- and E-carrier multiplexer or router rerouting
V.34bis modem backup

(DACS). This provides subscribers with the ability to reorganize, distribute and route individual circuits.

Bandwidth Management Service (BMS)

Bandwidth Management Service (BMS) represents an AT&T offering which enables subscribers to reallocate the routing of individual DS0 channels on a T1 circuit. This can be an extremely flexible tool if your organization has a fan-out network structure with a T1 line at one location used to terminate fractional T1 and 56/64 kbps leased lines routed to other locations.

10.2 THIRD PARTY AND USER-PROVIDED SOLUTIONS

There are several methods you can consider to obtain a digital circuit restoral capability via the use of a third party product or through the use of such equipment. Table 10.1 lists four commonly available circuit restoration methods you can consider.

VSAT

In examining the entries in Table 10.1 note that the use of a very small aperture terminal (VSAT) for satellite transmission will result in a transmission delay of approximately half a second. Although this delay can be compensated for by adjusting a transmission window for data transfer, it may be unsuitable for the backup of a circuit transporting voice for more than a short period of time.

The switched digital network

The use of the switched digital network can provide 56/64 kbps, 128 kbps, 384 kbps and 1.536 Mbps operating rate transmission capabilities. However, not all locations support those data rates, which means you will have to investigate the availability of a particular switched digital transmission capability based upon the locations supported by your organization.

T- and E-carrier multiplexer or router rerouting

The construction of a T- or E-carrier multiplexed or routed network enables subscribers to program equipment to compensate for circuit failures. An elementary example of circuit compensation is illustrated in Figure 10.2. In this example a circuit failure between locations A and B is compensated for by routing the data flow from A to B via C.

V.34bis modem backup

A V.34bis modem has an operating rate of 33.6 kbps over the switched telephone network. The effect of V.42bis compression can usually provide a data transfer capability as high as 115.2 kbps. Thus, several vendors offer a low speed digital circuit restoration capability by including V.34bis modems in their DSU/CSUs. Since backup occurs over the PSTN for which you are billed based only on usage, it may be more economical to use this method of circuit restoral than to use redundant or redundant diversity routed circuits.

REVIEW QUESTIONS

1 How can circuit redundancy provide a false sense of circuit restoral capability?

2 Describe the general end-to-end routing of a pair of route diversity circuits.

3 When should you consider route avoidance?

4 What level of circuit restoral capability is provided by route avoidance?

5 What is the difference between AT&T's automatic protection capability and the vendor's network protection capability?

6 What type of service permits a subscriber to modify the routing of individual DS0 channels?

7 Describe a problem associated with the use of VSAT for circuit backup.

8 Describe an effective method for backing up a low speed digital circuit.

INDEX

Index compiled by Geoffrey Jones

LOCAL AREA NETWORKING

PROTECTING LAN RESOURCES
A Comprehensive Guide to Securing, Protecting and Rebuilding a Network

With the evolution of distributed computing, security is now a key issue for network users. This comprehensive guide will provide network managers and users with a detailed knowledge of the techniques and tools they can use to secure their data against unauthorised users. Gil Held also provides guidance on how to prevent disasters such as self-corruption of data and computer viruses.
1995 0 471 95407 1

LOCAL AREA NETWORK PERFORMANCE
Issues and Answers
Second Edition

The performance of LANs depends upon a large number of variables, including the access method, the media and cable length, the bridging and the gateway methods. This revised text covers all these variables to enable the reader to select and design equipment for reliability and high performance.
1996 0 471 96926 5

LAN TESTING AND TROUBLESHOOTING
Reliability Tuning Techniques

Network testing is becoming a major requirement in corporate, industry and government computing. This book focuses on networking systems and the testing tools on the market today.
1996 0 471 95880 8

HIGH-SPEED NETWORKING WITH LAN SWITCHES

The demand for switching is on the increase as higher bandwidths are required from LANs, the internet and intranets. This book focuses on different types of LAN switches and how they fit in with current network devices.
1997 0 471 18444 6

VIRTUAL LANs
Construction, Implementation and Management

Virtual LANs allow network administrators to group users in a logical network rather than one based upon physical location. The book examines this new way of setting up networks from an intermediate level.
1997 0 471 17732 6